BASIC FACTS ON PHYSICS

BASIC FACTS ON PHYSICS

Aman Rao
M.Sc., Ph.D.

ANMOL PUBLICATIONS PVT LTD
New Delhi-110 002

ANMOL PUBLICATIONS PVT. LTD.
4374/4B, Ansari Road, Daryaganj
New Delhi-110 002

Basic Facts on Physics

First Edition 1995

Reprint 1999
ISBN 81-7488-079-8

PRINTED IN INDIA

Published by J.L. Kumar for Anmol Publications Pvt. Ltd., and
Printed at Mehra Offset Press, Delhi.

Preface

The book **Basic Facts on Physics** especially for the benefit of the students in undergraduate and postgraduate classes. It would be quite useful for all types of readers. If gives the meanings of various words in a simple and comprehensive manner. The words included in the book are those which we come across most frequently and so it is likely to improve the vocabulary and general awareness of the reader. The book has been written in a clear and iucid style. To provide more clarity to some terms diagrams have been provided alongwith them.

In compiling this book matter has drawn from different sources and the author is highly thankful to all such persons whose works have been consulted while preparing this book.

Comments on the book are most welcome.

Author

A

Aberration (1) A defect in a lens or mirror that causes blurring or distortion of the image. The three most important types are spherical aberration, chromatic aberration, and astigmatism. Spherical aberration is caused by rays from the outside of the lens or mirror being brought to a focus at a different point from those nearer to the centre. In chromatic aberration, different colours are focused at different points, since the refractive index of glass varies with the wavelength (2) An apparent displacement in the position of a star or other heavenly body due to the motion of the observer with the earth in its orbit round the sun.

Absolute Temperature (T). Temperature on thermodynamical scale based on Charle's law. According to Charle's law, the volume V of a gas at any tempeiature θ is related to its volume $V\theta$ at 0^0 by the relation

$$V = V_0 = \left(1+ \frac{\theta}{273} \right)$$

The absolute temperature is related to the temperature θ on the Celsius scale as

$$T = 273 + \theta$$

The temperature intervals on this scale are called degree absolute (^{0}A) or degree Kelvin (^{0}K).

Absolute Zero. Zero of thermodynamic temperature (0 kelvin). It is the lowest temperature theoretically attainable. At this temperature the kinetic energy of atoms and molecules is minimal. It is equivalent to - 273.15^{0}C or - 459.67^{0}F.

Acceleration. The rate of change of a body's velocity. Linear acceleration is the rate of change of linear velocity. It is measured in such units as metres per second per second. Angular acceleration is the rate of change of angular velocity and is measured in such units as radians per second per second.

Acceleration of Free Fall (g). Formerly called acceleration due to gravity; the acceleration of a falling body when air resistance is neglected. Caused by gravitational attraction between the body and the earth, it varies slightly at different points on the earth's surface. Its standard value is 9.806 metres per second per second.

Accelerator Principle. The economic principle that investment will accentuate economic booms and *depressions*. As income rises, businesses gain confidence in the expected future level of demand and increase their investment in plant and equipment; this pushes up employment in the capital-goods industries and heightens the boom. The converse applies to a slump.

Accelerators. Large machines that are used for accelerating beams of charged particles (electrons, protons, etc.) to very high speeds primarily for research in particle physics. The particles are accelerated by electric fields either in a straight line, as in the *linear* accelerator, or in a circle, as in the *cyelotron synchrotron* and *synchrocyclotron.* The beam is confined to its path by magnetic fields. The energies of the particles are measured in millions of electronvolts (MeV) or giga-electronvolts (1000 MeV = 1 GeV), some modern accelerators attaining several hundred GeV. Particle accelerators are used by directing a beam of particles at a stationary target or, for greater energy, by colliding two beams of particles together. Accelerators are also used to create artificial isotopes and in *radiotherapy*. The first accelerator was a linear accelerator, produced in 1932 by *Cockcroft* and *Walton*.

A Chromatic Lens. A combination of lenses used to eliminate chromatic *aberration* in an optical system. The simplest type has two lenses of different powers made from different kinds of glass. The chromatic aberration of one lens is cancelled by the chromatic aberration of the other lens.

Acoustics. The branch of physics concerned with the production, propagation, reception, properties, and uses of sound. It has several subdivisions. The most important, architectural acoustics, is concerned with the design of public auditoriums so that sounds can be heard in all parts of them with the maximum clarity and the minimum distortion. *Ultrasonics* is the study of very high frequency sound. The structure and function of sound sources, such as loudspeakers, and sound receptors, such as microphones, also form part of acoustics. Other fields include speech communication and the design of machines that can understand spoken instructions.

Adiabatic Demagnetisation. A process wherein a sample of a paramagnetic salt is cooled in liquid helium while in a strong magnetic field; the sample is then demagnetized by switching off the field, this results in cooling the substance further. The process is used to produce very low temperature close to 0^0K.

Aerial (*or antenna*). An electrical conductor that transmits and receives radio or other electromagnetic waves. An oscillating *electromagnetic* field, caused by waves from a distant source, induces an oscillating current in the receiving aerial. A transmitting aerial works by the same process in reverse, creating waves from an electrical signal. A modern aerial for UHF (ultra-high frequency) and VHF (very high frequency) consists of a dipole formed from two metal rods, each approximately one-quarter of the operating wavelength. In the Yagi aerial named after H. Yagi (1886-1976), a reflector rod is set behind the dipole and several director rods are placed in front of it. This provides a more directional array than the simple dipole and is widely used as a receiving and transmitting aerial for television.

Aeronautics. The science and history (*see also* aircraft) of flight. An object flying through air is subject to four basic forces: its own weight (vertically downwards as a result of gravity), lift (to counterbalance its weight and keep it in the air), thrust (to force it through the air), and drag (resulting from friction between the body and the air). Birds and insects use their wings to provide both lift and thrust; man, in his heavier-than-air fixed-wing craft, uses an aerofoil to provide the lift and an *internal-combustion* engine (propeller or jet) to provide the thrust *Helicopters* use rotating aerofoils to provide lift, while *rockets* use no lift surfaces, the jet of expanding gas providing both lift and thrust. Man's lighter-than-air craft use helium or hydrogen to reduce the craft's weight in relation to the volume of air it displaces to such an extent that the lift is provided by buoyancy. The use of aerofoils as lift surfaces depends on *Bernoulli's* theorem, according to which the total energy of a flowing fluid remains constant; thus, if the velocity of the fluid increases, its pressure decreases in proportion. An aerofoil is a wing so shaped that (at subsonic speeds) air is accelerated over its rounded leading edge and curved upper surface, causing a reduced pressure above it. A smaller reduction in air velocity on its underside causes a slightly increased pressure below it. The combination of these pressure differences provides the lift. The design of practical aircraft wings has to take into account a number of complex factors, including suitable streamlining to avoid *turbulence* in the airflow, stability over different angles of attack, the provision of suitable control surfaces (flaps, ailerons, etc.) and adequate strength and rigidity. A supersonic speeds these forces are somewhat altered and the aerofoil has to be more sweptback and more streamlined. At hypersonic speeds (i.e., in excess of five times the speed of sound) the *aerodynamics* changes again and blunter noses and even smaller wings are needed.

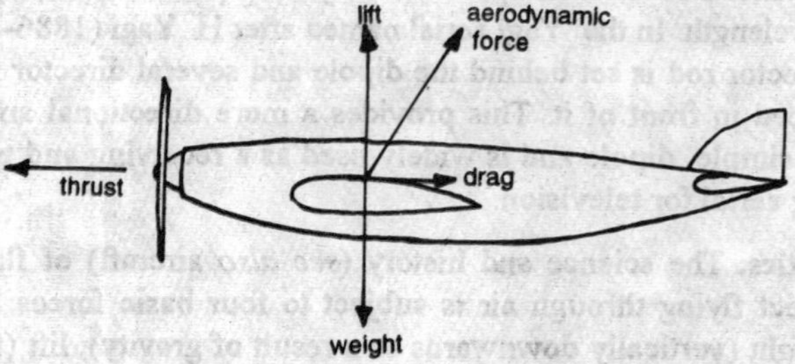

forces acting on aircraft

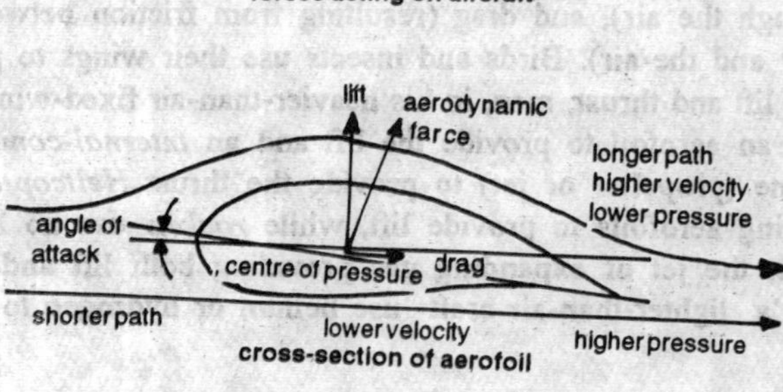

cross-section of aerofoil

Aeronautics. The forces acting on an aircraft. The aerofoil cross-section shows how the lift, which keeps it in the air, results from the passage of the aerofoil through the air, causing a lower pressure above it and a higher pressure below it.

Aether. An imaginary fluid which was thought to permeate all space through which light waves were assumed to propagate. It was shown by Michelson-Morley experiment that the existence of such a medium is unfounded.

Aircraft. Any machine capable of flying. Man's attempts to fly fall into two categories: those using lighter-than-air machines and those using heavier-than-air machines. The latter include wing-flapping birdlike devices, rotating wing machines, and fixed-wing aircraft, first *gliders* and then powered aeroplanes. By the end of the 19th century it was clear to all but the most eccentric that man was both too weak and too heavy to emulate birds. Flapping wings combine both thrust and lift in one device; man needs to separate these two components, obtaining lift from a fixed wing and thrust from an engine. 19th-century experience of gliding, especially by Otto *Lilienthal,* provided the Wright Brothers with the information they needed to build their first powered aircaft. The power source was provided by the Otto-Daimler *internal-combustion* engine. By 1907 the Wrights were able to remain airborne for 45 minutes; in 1909 the Frenchman Louis *Bleriot* flew across the English Channel, and in the same year the French rotary Gnome engine revolutionized aircraft-engine design (this engine and its derivatives powered many early aircraft, including several used in World War I). By the beginning of the war aircraft were sufficiently advanced to be used for reconnaissance and their usefulness as bombers soon became evident. Fast manoeuvrable fighters to shoot down the slower heavily laden bombers were an obvious subsequent developement. By the end of the war aerial combat was established as an integral part of modern warfare, both sides being equipped with a range of fighters, bombers and reconnaissance aircraft.

After the war and during the 1920s, air circuses and flying clubs run by ex-wartime pilots using World War I aircraft sprang up all over the world, popularizing the concept of flying and heralding the age of civil aviation. In 1924 Imperial Airways was formed in the UK, using Handley *Page airliners,* and during the 1930s a worldwide network of commercial routes developed. The Atlantic was first flown nonstop from New York to Paris in 1927 by Charles *Lindbergh* and by 1939 there was a transatlantic service using Class C Short flying boats with inflight refuelling.

Between the wars all the main countries of the world were also building up their air forces. By the outbreak of World War II aircraft of all kinds were poised for aerial combat. The first RAF bombing raid of the war was carried out by Vickers Wellingtons and Bristol Blenheims (of which some 3500 were built). During the course of the war these bomers were augmented by some 6000 Handley Page Halifaxes (which entered service in 1940) and 7400 Avro Lancasters (1942). The USA's main bombing force constisted of 12 700 Boeing B 17s , 18 000 Consolidated B24 Liberators, and towards the end of the war, a fleet of Boeing B29 Superforteresses. Germany entered the war with 370 Dornier Do 17s and the Heinkel He 111, some 7300 of which had been made by the end of the war. The Junkers 88 entered service in 1940 and at the end of the war (1945) the Arado Ar 234 jet bomber was used for the first time. Dominant British fighters were the Supermarine Spitfire and the Hawker Hurricane, the latter being slightly slowere than the main German fighter, the Messerschmitt Me Bf109. Later British fighters included the Bristol Beaufighter and the De Havilland Mosquito. The US fighter pilots had some 10000 Lockheed P38 Lightnings and 15000 P51 Mustangs. The British jet-powered Gloster Meteor entered service in 1944, as did the German Messerschmitt Me 262 jet.

By the end of the war the practicality of the *jet* engine had been established and it has dominated aircraft design ever since. However, the first postwar generation of civil aircraft used the jet engine to drive propellers (e.g. the Vickers Viscounts), the first true jet to enter passenger service being the ill-fated British Comet. The Boeing 707 (with its four engines in pods suspended below the wins) followed in 1954 and the French Caravelle (with two rear-mounted engines) in 1959.

The first aircraft to break the *sound* barrier were military and nearly all modern warplanes are supersonic and armed with missiles. Examples include the British Hawker Hunter and Avro Vulcan; the US McDonnell Douglas Phantom, Convair Hustler, and General Dynamics Swingwing F111; the French Mirage; and the Soviet MiGs. The latest generation of warplanes includes the US SR-71A and B2 Stealth bomber. The first propeller-drivern fighter plane to be developed since the advent of the supersonic jet fighter was the 'heliplane' P1233-1 (1987), which is designed for manoeuvrability in low-level dogfights. The first supersonic passenger aircraft (SST) to fly was the Soviet Tupolev Tu-144 in 1968. This was followed a year later by the Anglo-French Concorde, which despite opposition from conservationists, is now in serevice. However, long-and medium-range passenger services are likely to be dominated for most of the remainder of this century by the wide-bodied (jumbo) jets, such as the Boeing 747 and the European Airbus.

Aircraft Carrier. A navel vessel with a large flat deck for launching and landing warplanes. The first fight from the deck of a ship was made in 1910, and the first true aircraft carrier, HMS *Argus,* was completed for the Royal Navy in 1918, too late for action in World War I. Carriers played a dominant role in World War II, despite early predictions that *battleships* would be the most important warships. Carriers were especially effective in the war against the Japanese, where they were employed in combined UK and US operations and were instrumental in destroying the Japanese fleet. After World War II carriers came to be regarded chiefly as tactical units, although they saw considerable action in the Korean and Vietnam Wars. The USS *Enterprise,* the first nuclear-powered carrier (1961), displaced 76 000 tonnes and steamed more than 432 000 km (270 000 mi) before requiring refuelling. The introduction of large and heavier aircraft required the refitting of carriers. Because of the increased range of aircraft, the enormous cost of carriers, and the developoment of sophisticated missiles, there have been few aircrat carriers build in the 1970s, and indications are that their size, cost, low speed, and vulnerability will preclude their extensive use in the future. HMS *Ark Royal,* the UK's last aircraft carrier, was withdrawn from service in 1979, but a new *Ark Royal* was commissioned in 1985.

Airships. Dirigible *balloons* that obtain their thrust from a propeller. The first airship to fly was a French steampowered machine, designed in 1852 by H. Giffard; however, the first practical airship was the electrically powered *La France* (1884), built by Renard and Krebs. By 1900 the initiative in airships had passed to Germany, with the machines of Count Ferdinand von Zeppelin (1838-1917) leading the field. Between 1910 and 1914 Zeppelins were in extensive passenger service, carrying some 35000 passengers, without mishap. In World War I these machines were used by the Germans to bomb England—the first effective use of aerial bombardment. From then onwards the history of airships is one of disaster followed by disaster. The British *R101* caught fire at Beauvais in 1930, the US *Shenandoah* and the *Akron* were lost in 1933, and the German *Hindenburg* was destroyed in 1937. These disasters with *hydrogen*-filled airships cost many lives and gave them a reputation from which they have never recovered. However, the availability of the non-flammable gas helium has created a mild revival in recent years.

Albedo. A measure of the reflecting power of a non-luminous object, such as a planet or natural satellite or a surface feature on such a body. It is the ratio of the amount of light reflected in all directions from the object to the amount of incident light. Clouds, snow, and ice have high albedos while volcanic rocks have very low albedos.

Alternating Current (ac). Electrical current that periodically reverses its direction. It is the form of current that is produced when a coil of wire rotates in a magnetic field and, as this is the way in which current is produced in *power stations* by *electric generators*, it is the form of current most widely used. The electromotive force (emf), E, produced by a generator is equal to $E^1 \sin \omega t$, where E^1 is the maximum emf, ω is the angular velocity of rotation, and t is the time. Thus the current has the form of a sine wave, with a frequency $\omega/2\pi$

The chief advantage of ac is that the voltage can be stepped up with a transformer before transmission to minimize energy losses in the lines and then reduced to a safer level by another transformer for domestic use. In the UK the ac supply has a frequency of 50 hertz, is transmitted at several hundreds of kilovolts, and is used at 240 volts. In the USA the supply frequency is 60 hertz.

Alternative Energy. The investigation and use of new sources of power, based on natural energy flows in the environment. Most countries now rely heavily on *fossil fuels* (oil, coal and natural gas) and nuclear power for their *energy* needs. However, reserves of fossil fuels are declining and, although dates for their total exhaustion vary, their price must rise as they become increasingly scarce. They also contribute to the *greenhouse* effect by emitting carbon dioxide into the atmosphere. The estimates of reserves of recoverable uranium are also uncertain and there is opposition to the development of nuclear breeder reactors, mainly on the grounds of safety and environmental hazard from waste disposal. Fusion reactors are still very much in the experimental stage. Although nuclear power is still regarded as the main option, it is also necessary to explore the alternatives. Renewable energy sources are those that do not use up finite mineral resources. Of these *solar power, wind power* and *tidal power* are being most seriously investigated in the UK. *Hydroelectric* power is already in use and has limited potential for further expansion.

Geothermal power, waver power and *biomass energy* are also being actively explored. It is estimated that 20-25% of the UK's energy requirements could be met by renewable sources by 2025.

Amagat. A unit of volume often used in the study of the equation of state of gases. By definition the molar volume of a gas at 0^0C and 1 atmosphere is 1 amagat. The exact value of the unit depends upon the gas considered, but is approximately 22.4 liters $mole^{-1}$. There is also an amagat unit of density equal to the density of a gas in which 1 mole occupies a volume of 1 amagat.

Ammeter. An instrument for measuring electric current. The two most common types are the moving-coil and the moving-iron ammeters. The moving-coil ammeter is more sensitive but will only measure alternating current. The moving-iron ammeter will measure both alternating and direct current but is less sensitive and its scale is non-linear. Some instruments are electronic and have a digital display.

Amonton's Law. In any gas whose volume and mass are kept constant, the same rise in temperature produces the same increase of pressure.

Ampere (A). The SI unit of electric current equal to the current that when passed through two parallel infinitely long conductors placed 1 metre apart in a vacuum produces a force between them of 2×10^{-7} newton per metre. This 1948 definition replaced all former definitions. Named after A.M. *Ampere.*

Ampere's Law. The strength of the magnetic field at any point produced by a current (I) flowing through a conductor of length is proportional to Il/d^2 where d is the distance between the point and the conductor. Named after A.M. *Ampere.*

Amplification Factor (µ). Refers to the ratio of the change in plate voltage that can produce a change in the plate current to the change in grid voltage that produces the same change in plate current for a triode valve.

Amplifier. A device for increasing the magnitude of some quantity by using power from an external source, especially an elecronic device for intensifying a signal in an alternating-current circuit, with an external steady voltage. Amplifiers are designed to multiply the input (current, voltage, or power) by a specific factor, known as the gain and often consist of several stages the output from one stage becoming the input to the next stage. This method is used in the specialized type of amplifier used in sound-reproduction systems.

Amplitude. Refers to the maximum displacement of an oscillating quantity from its means position. For a simple pendulum, the amplitude is equal to the distance between the mean and extreme position of the bob.

Amplitude Modulation. A method of transmitting information by radio waves. In this method, the amplitude of the carrier waves is changed (or modulated) in accordance with the frequency of the signal to be transmitted.

(a)

(b)

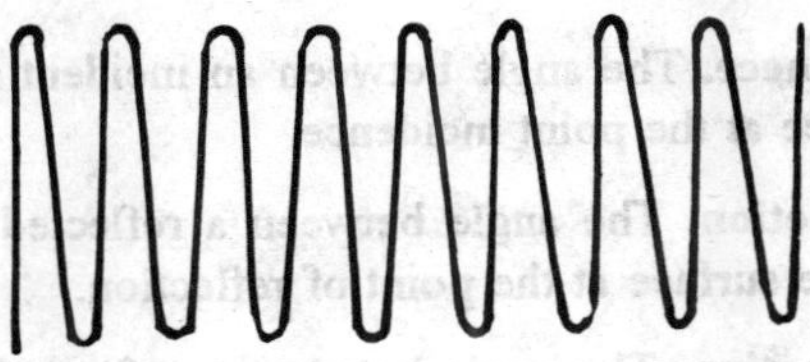

(c)

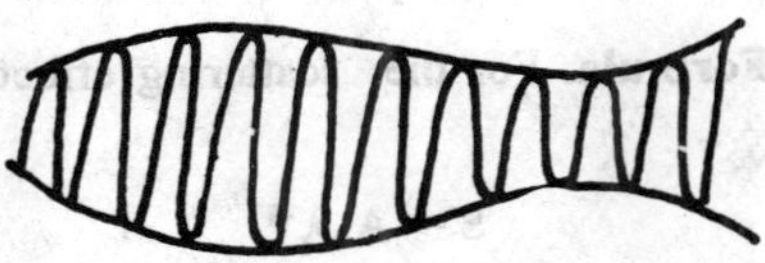

Anastigmatic Lens. Refers to a lens in which astigmatism is reduced to a minimum. Such a lens has different curvatures in different direction.

Anderson Localization. For understanding the theory of electrons in certain non-crystalline media, P.W. Anderson has proposed the use of a potential-energy function which is non-periodic by the addition of a random potential energy to the periodic frunction used for crystalline solids. If the ratio of this random potential energy to the energy bandwidth associated with the original periodic function is higher than approximately 5 (the ratio depends slightly on the coordination number associate with the material), the wave-functions for the electrons decay exponentially with distance, thus giving rise to electron localization.

Andrade's Creep Law. Andrade showed that when a load is applied at the beginning of a creep test, the instaneous elastic elongation is followed by a transient state in which strain varies as $t^{1/3}$ and, finally, a steady state is reached in which there is a constant rate of creep under constant effective stress.

Anemometer. An instrument for measuring the velocity of a fluid, often the velocity of the wind. In one type, the fluid drives a small windmill or set of cups, the rate of rotation of which is a measure of the velocity of the fluid. Two other common types of anemometer are the *Pitot* tube and the *Venturi* tube.

Angle of Friction. Refers to the angle between surfaces at which slipping first (just) occurs. It depends on the nature of the surfaces that are in contact.

Angle of Incidnece. The angle between an incident ray and normal to the surface at the point incidence.

Angle of Reflection. The angle between a reflected ray and the normal to the surface at the point of reflection.

Angle of Refraction. The angle between a refracted ray and the normal to the surface at the point of refraction.

Angstrom's Formula. For the scattering effect of dust in the atmosphere

$$S = A\,\lambda^{-B}$$

where λ is the wavelength, B depends on the particle size and A is a constant.

Angular Acceleration. It is the rate of change of the angular velocity of an object. If θ is the angular displacment of an object, the angular acceleration (α) is,

$$\alpha = \frac{d^2\theta}{dt^2} = \frac{d\omega}{dt}$$

where $$\omega = \frac{d\theta}{dt}$$

is its angular velocity.

Angular Displacement (θ). Refers to the angular or rotational displacement of an object from a given position about an axis.

Angular Frequencey (ω) . It is the number of complete rotations made per unit time.

The angualr frequency w of a simple harmonic motion of frequency f is given by $\omega = 2\pi f$. Its unit is hertz (Hz).

Angular Magnification. Refers to magnifying power. The ratio of the angle subtended by the image formed by an optical device at the eye to the angle subtended by the object. In case of a mocroscipe, the angle substended by the object is taken to be when the object is placed at the least distance of the distinct vision, viz., 25 cm from the eye.

Anomalous Dispersion. A sudden discontinuity in the graph of μ (refractive index) versus λ (the wavelength of light) which is due to the high absorptivity of the medium at wavelengths that lie in the region of absorption band.

Anomalous Expansion. Expansion of certain substances with a decrease in temperature. Such a behaviour is shown, by water when it is cooled below 4^0C. Water expands on cooling between 4^0C and 0^0C.

Antiferromagnetism: The anti parallel alignment of succesive magnetic moments in a substance. The susceptibility of such susbstances increases with rise in temperature upto a certain point, called Neel temeprature, above which the substance becomes paramagnetic.

Antimatter. Hypothetical matter in which the constituent atoms consist of antiparticles. For every elementary particle there exist an antiparticle that is identical except for centain of its properties, such as electric charge and isospin number, which are of equal magnitude but opposite in sign. The photon and the neutral pion are their own antiparticles. An atom of antimatter would contain a nucleus of antiprotons and antineutrons surrounded by positrons (antielectrons). If matter were to meet antimatter, they would annihilate each other in a burst of radiation.

Antinodes. These are the points of maximum displacement in a stationary wave pattern. The distance between successive antinodes in a stationary wave is equal to one half of the wavelength.

Anti Particle. Corresponding to every particle, is supposed to exist another particle which has the same mass and spin but opposite charge. When it comes in contact with its corresponding particle, the two annihilate to produce radiation.

Antonie Equation. An expression for the vapour pressure *p* of a condensed solid or liquid as a function of absolute temperature T:

$$\ln p = A - \frac{B}{C+T}$$

where A, B and C are constants obtained experimentally.

Antonoff's Rule. The interfacial surface tension γ_{AB} between two saturated liquid layers A and B, in equilibrium, is equal to the difference between the surface tensions against vapour or air of the two mutually saturated solutions, i.e.

$$\gamma_{AB} = \gamma_A - \gamma_B$$

Aperture. The size of the opening that admit light in an optical instrument. The effective diameter of mirror or lens.

Aplantic. Refers to a reflecting or refracting surface which is aplantic wiht respect to two points A and B, if it produces a point image at B of an object placed at A, irrespective of the angle at which the rays fall on the surface from the object.

Aplanatic Lens. A lens that minimize both its astimatic and coma aberration.

Apochromatic Lens. A lens designed to correct for chromatic aberration at three different wavelengths. Such lenses are constructed of three or more types of glass. They have better correction than achromatic lenses, which correct at two different wavelengths (usually in the red and blue regions of the spectrum).

Apogee. Refers to the greates distance of a satelite from the earth during its orbit around the earth.

Aparent Depth. The depth of a vessel having water or a pond appears to be less than the actual depth. The depth as it appears to the eye is known as apparent depth and is equal to Real Depth/μ, where μ is the refractive index of water.

Apparent Expansion. It is equal to the difference between the actual expansion of a liquid and the expansion of the vessel containing it.

Aqueous Humour. The liquid present between the cornea and the lens in the eye.

Archimedes. (c.287-c.212 BC) Greek mathematician and inventor, regarded as the greatest scientist of classical times. Archimedes was born in Syracuse, Sicily, and studied in Alexandria, afterwards returning to Syracuse, where he remained for the rest of his life. He is best known for his discovery of *Archimedes'* principle, supposedly in response to the King of Syracuse asking him to determine whether a gold crown had been adulterated with silver. Legend has it that he made his discovery while taking a bath and ran through the streets of Syracuse shouting "Eureka!" He is credited with inventing *Archimedes'* screw, although the device was probably already known to the Egyptians. He was killed during the Roman invasion of Syracuse.

Archimedes' Principle. The principle that when a body is partly or wholly immersed in a fluid its apparent loss of weight is equal to the weight of the liquid displace. Named after *Archmedes*.

Archimedes' Screw. A device for raising water, reputed to have been invented by *Archimedes*. It consists of an inclined helical screw rotated about a central axis in a trough of water.

Armature. (1) Refers to the coils (s) of a dynamo or an electric motor. (2) A part of an electric machine wherein an e.m.f. is induced by a changing magnetic field, e.g. electromagnetic relay.

Armstrong, Edwin Howard (1890-1954). US electrical engineer, who in 1939 introduced frequency modulation as a method of broadcasting radio signals to reduce the static associated with amplitude modulation. He had earlier invented the *superheterodyne* receiver, which improved radio reception are greatly increased the popularity of radio sets. A contentious and litigious man, he committed suicide.

Aspirator. A device used to produce a partial vacuum for high vacuum pumps. It is based on Bernoulli's principle and reduces the pressure in a vessel in a very short duration to about 2 cm of mercury.

Astatic Coils. An arrangement of two coils which are used in insensitive electrical instruments, so that the resultant external magnetic field produced by them due to an electric current passing through them, and the e.m.f. induced in them by an external magnetic field are zero.

Astigmatism. A form of *aberration* that can occur in mirrors and lenses (including the eye). It results when the curvature is different in two mutually perpendicular planes; rays in one plane may then be in focus while rays in the other are out of focus. It is corrected in the eye by the use of cylindrical lenses.

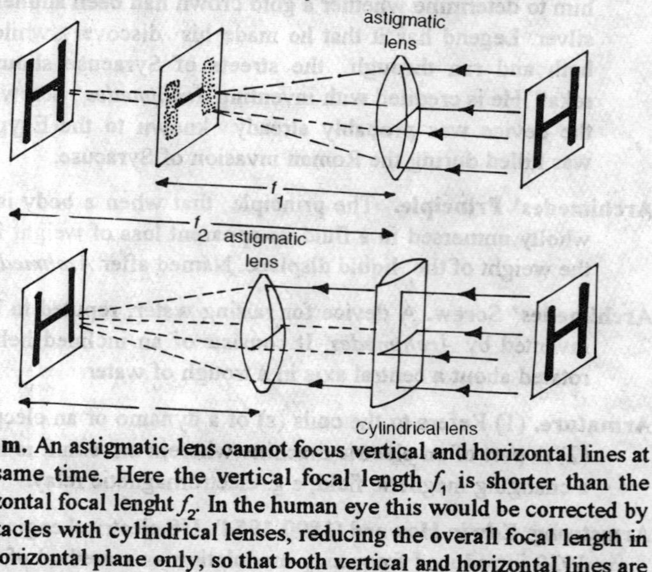

Astigmatism. An astigmatic lens cannot focus vertical and horizontal lines at the same time. Here the vertical focal length f_1 is shorter than the horizontal focal lenght f_2. In the human eye this would be corrected by spectacles with cylindrical lenses, reducing the overall focal length in the horizontal plane only, so that both vertical and horizontal lines are in sharp focus at distance f_1.

Astronomical Unit. (AU) A unit of length equal to the mean distance between the earth and the sun (1.495×10^{11} meters, 92.9×10^6 miles).

Astrophysics. The study of the physical and chemical processes and characteristics associated with celestial object. It is based on theories developed in astronomy, physics, and chemistry and on observations of the radiation emitted by the objects. Originally only studies at optical and then at radio and infrared wavelengths were made, but sources of X-rays, gamma rays, and ultraviolet radiation can now also be observed.

Atkinson Cycle. A working cycle for an internal combustion engine where the expansion ratio exceeds the compression ratio. It is more efficient theoretically than the *Otto Cycle* but is difficult to achieve in practice.

Atmosphere (Unit). A unit of pressure equal to 101 325 pascals or 760 millimetres of mercury.

Atmospheric Pressure. The pressure exerted by the atmosphere. It decreases with altitude and, at groud level, varies around 760 millimetres of mercury, 101 325 pascals, or 1013.25 millibars.

Atomic Clock. A highly accurate clock based on the frequency at which certain atoms or molecules vibrate between two states. For example, the nirtogen atom in the ammonia molecule vibrates through the plane of the three hydrogen atoms and back again with a frequency of 23 870 hertz. In the ammonia clock, a quartz-crystal oscillator feeds energy into ammonia gas at this frequency. The ammonia only absorbs energy at this frequency, enabling a feedback circuitto controlthe frequency of the oscillator.

Atomic Heat. The product of atomic weight and specific heat of an element. The atomic heat of all solid elements is 6.4 calorie/gm^{0}K. at ordinary temperatures. This is called Dulong and Petit's Law. At low temperatures, the atomic heat decreases with temperatures, approaching zero as temperature goes to zero kelvin.

Atomism. The philosophical attitude that seeks irreducible elements, whether of matter or thought, to account for the (compound) phenomena that we actually experience. The Greek *Democritus* believed that the world is composed of qualitatively similar atoms of different shapes. In this view, atoms may be either completely independent of one another or related in contingent or necessary ways. *Lucretius* was an influential proponent of atomism, and *Gassendi, Boyle,* and *Locke* revived it in the 17th century. Logical atomism, held temporarily by both *Russell* and *Wittgenstein,* assumed that there were unanalysable specks of meaning that could be articulated in atomic propositions, could be derived.

Attenuation. The reduction in amplitude of an electromagnetic wave or an electric current during transmission. In electric circuits attenuation is often introduced to reduce unwanted components of a signal. Broadcast electromagnetic radiation is attenuated as it passes through building, etc., and to a lesser extent by its passage through the atmosphere. In space, however, the attenuation is negligible.

Attenuation Equalizer. An electrical network used to provide compensation for attenuation distortion in a specific band frequencies.

Audibility, Limits of. The limits of frequencies that can be heard by the human ear. The lowest audible frequency is about 20 Hz, and the highest frequency is 20,000 Hz.

Audio Frequency A frequency in the range 20 to 20 000 hertz, i.e., the range of frequencies to which the human ear is sensitive.

Auger Effect. When a K electron is ejected from an atom, an outer electron can fall into the K shell and give up its energy in two ways.

I. By radiation which gives rise to X-ray emission lines.

II. By transferring its energy to one of the outer electrons which is ejected from the atom. This is known as the Auger effect.

Aurora. A display of diffuse changing coloured light seen high in the earth's atmosphere, often taking the form of streamers or drapery and usually green or red. Aurorae occur predominantly in polar regions when energetic charged particles from the sun become trapped in the earth's magnetic field. The rapidly moving particles interact with atoms and molecules in the upper atmosphere and cause them to emit light. In the N hemisphere they are called the Northern Lights.

Automatic Gain Control. (a.g.c.). A method that automatically holds the output volume constant in a radio receiver, despite changes in the input signal.

Automatic Pilot. A device utilizing a gyroscope for keeping an aircraft on a given course. When the aircraft goes off course, the gyro axis rotates, operating elecrical contacts, which make the necessary adjustments to the control surfaces.

Autoradiograph. If a thin specimen of metal, biological tissue, etc., is labelled with a radioisotope and placed in contact with a photographic plate for a suitable exposure time an image is produced by the action of the radiation emitted. On developing the film, called an autoradiograph, the distribution of the radio-isotope in the specimen can be seen.

Autotransductor. A transductor wherein the same windings are used for the main and control curents.

Autotransformer. A transformer having a single windings tapped at intervals, instead of two or more independent windings. The voltage drop across each tapped section is related to the total applied voltage in the same proportion as the number of turns of the section is related to the total number of turns of the winding.

AWACS (Airborne Warning and Control System). A US *radar* system mounted on an aircraft to give early warning of the movements of enemy aircraft and, if appropriate, to direct intervention measures against them.

Axial Ratio. The relative lengths of the three edges of the unit cell of a crystal lattice, taking that of the b axis as unity.

Axis (of a Lens). For spherical lenses refers to a straight line perpendicular to each face and joining the centres of the two surfaces. For cylindrical and sphero-cylindrical lenses, refers to a line (on the surface) parallel to the geometrical axis of the cylinder surface.

Azbel-Kaner Resonance. A type of cyclotron resonance in high-purity metals as liquid helium temperatures, first observed by Azbel and Kaner in 1957. A magnetic field is applied parallel to the surface of the metal in the same direction as an oscillating electric field. Current can only flow in a region next to the surface having a depth called 'anamalous skin depth'. The radius of the electron orbits under the above conditions is much larger than this skin depth, so electrons can only be accelerated once per cycle and only if their orbits enter the skin depth region. The frequency of the electric field must be the cyclotron frequency or an integral multiple of it.

B

Background Radiation. Low-intensity radiation naturally present on the earth. It results either from the bombardment of the earth by *cosmic rays* or from naturally occurring radioactive substances in the earth's crust.

Balance. A sensitive device for comparing two masses, consisting of a beam pivoted at its centre (usually on an agate knife edge) with pans hanging from each of its ends. The material of unknown mass is placed.in one pan and standard weights are placed in the other. A pointer indicates when the beam is horizontal and the whole device is enclosed in a glass case to avoid draughts and temperature changes. The accuracy of a balance is increased by using a rider—a small weight hung on a calibrated scale on the balance arm itself. A standard balance will weigh to the nearest 0.0001 g, while sensitive *microbalances* can weigh objects of 1 microgram.

Ballistic Galvanomenter. A galvanomenter that is used for measuring the total amount of charge that passes through a circuit due to a momentary current. Its period of oscillation is large compared to the time for which the current flows through it. The miximum deflection produced in such a case is proportional to the total charge passing through it.

Ballistic Missiles. Rocket-powered nuclear missiles without wings or other lift surfaces that are propelled to desired altitudes and velocities and then follow an unpowered trajectory similar to that of a projectile fired from a gun. Their accuracy requires careful thrust calculations and on-board preset or inertial guidance systems and is calculated as a probability (e.g. 45-60%) of landing within a stated radius about their target (Circular Error of Probability). Intercontinental ballistic missiles (ICBMs) are capable of reaching any point on the surface of the earth. Both the USA and the Soviet Union possess large numbers of these, some of which (MIRVs-multiple independently targeted re-entry vehicles) have up to ten separate warheads.

Ballistic Pendulum. A heavy pendulum, that can be used to measure the velocity of a projectile. It is struck by the projectile whose velocity can be calculated by measuring the resultant displacement of the pendulum.

Ballistics. The study of projectiles and the extent to which their trajectories are affected by shape, propulsion systems, gravity, temperature, wind, etc. There are three branches: interior, dealing with all aspects of propulsion within a gun barrel or at launch; exterior, concerned with the trajectory of the projectile in flight; and terminal, relating to the effects of the missile on the target.

Balloons. Lighter-than-air craft, consisting of a bag of gas that displaces a volume of air of greater mass than the total mass of the balloon and its contents. The first successful balloon flight, indeed man's first aerial voyage, was made in 1783 by the *Montgolfier* brothers' hot-air balloon; it flew 9 km across Paris. Two years later a *Charliere*, hydrogen balloon (designed by J.A.C.Charles the formulator of *Charles's* Law) flew across the English Channel. In 1821 coal gas was used for the first time as a cheap alternative to hydrogen. This opened the way for many exploits by showmen, scientists, and explorers. Balloons must be tethered (for parachuting, scientific experiments, etc.) or they will fly where the wind blows them; dirigible balloons (*see* airships) first appered in the middle of the 19th cnetury. However, by the end of the century interest in flying was centred on heavier-than-air machines.

The use of balloons in war began with Napoleon's observation balloons in 1794, after which they continued to play a sporadic but largely ineffectual part, until their extensive use in both World Wars in the form of barrage balloons.

The sport of *ballooning* has enjoyed a revival in recent years, the preferred vehicle being a hot-air balloon carrying its own propane air heater. The height record for a manned balloon is 30 480 m (D. Simons, US Air Force, 1957). The first Atlantic crossing by a helium balloon was achieved in 1978. In

1987 Richard Branson made the first crossing by hot-air balloon.

Bar. Unit of pressure in the C.G.S. system. It is equal to the pressure exerted by a mercury column 76 cm long and its value is 10^6 dyne/cm^2.

Bardeen-Cooper-Schrieffer (BCS) Theory of Superconductivity. A quantum theory of superconductivity in which interation between pairs of electrons gives, for the electron system, a ground state which is separated from the excited states by an energy gap. The presence of this energy gap accounts for the normal/superconduting phase transition and the critical magnetic field which leads to the destruction of superconductivity. The theory accounts for the *London Penetration Depth* and the *Pippard Coherence Length*.

Barkhausen Effect. When a ferromagnetic material is magnetized in an increasing magnetic field, the magnetization does not increase smoothly with field. This is because of irregular fluctuations in the motion of the domain walls which separate regions of differeing directions of magnetization in the material.

Barkhausen-Kurtz Oscillations. Oscillations occuring in an electronic valve, wherein electrons of high velocity are expelled from the filament, meet the repelling effect of the grid-anode field, stop, return through the grid, and so on, eventually landing on the grid wires. The current accompanying this motion may be used to produce oscillations of frequencies between 300 and 1500 MHz.

Barlow Lens. A planoconcave lens placed between the objective and eyepiece in a telescope. It increase magnification.

Barn. Symbol: *b*. A unit of nuclear cross section. 1 barn = 10^{-28} metre.

Barnet Effect. A long iron cylinder rotating at high speed about its longitudinal axis, develops a slight magnetization propotional to the angular speed of rotation. This magnetization is on account of the effect of the rotation on the electornic orbits. In the atoms of the iron and on the electrons themselves, which have their won intrinsic spin.

Barograph. A recording barometer. It consists of an aneroid barometer operating a pen which traces a line on a sheet of graph paper mounted on a slowly revolving drum.

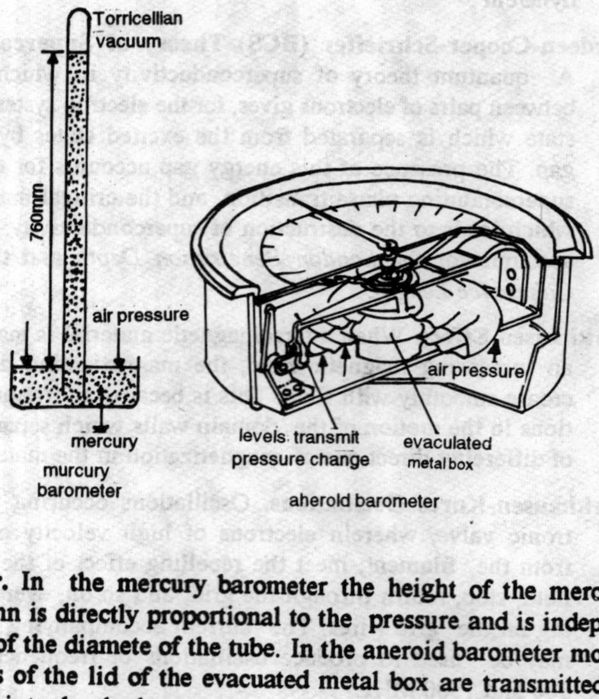

Barometer. In the mercury barometer the height of the mercury column is directly proportional to the pressure and is independent of the diamete of the tube. In the aneroid barometer movements of the lid of the evacuated metal box are transmitted to the pointer by the levers.

Barometer. An instrument for measuring atmospheric pressure. There are two main types: the mercury barometer and the aneroid barometer. In the mercury barometer, atmospheric pressure forces mercury from a reservoir into a vertical evacuated glass tube marked with a scale. The height of the mercury column is directly proportional to the atmospheric pressure. In the aneroid barometer, variations in the atmospheric pressure on the lid of an evacuated metal box cause a pointer to move round a dial. The aneroid barometer is less sensitive than the mercury barometer but it is smaller, more portable, and more convenient to use.

Baryon. A collective term for *nucleons* and other elementary particles that have a proton or neutron in their decay products. All baryons have a *quantum* number called the baryon number of +1 (or-1 for antiparticles), which is conserved in interactions.

Battery. A number of electric cells joined together in parallel. The common dry battery consists of one or more *Leclanche* cells. The alkaline dry battery has a negative electrode of zinc, a positive electrode of manganese dioxide, and an electrolyte of potassium hydroxide. A mercury dry cell is the same except that the negative electrode is mercury oxide mixed with graphite.

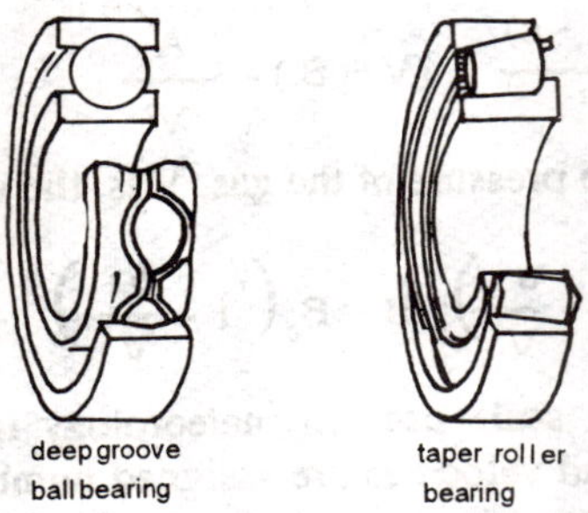

Bearings. A support for a rotating shaft or the interface between a crank and a reciprocating part. Bearings are designed to reduce friction and wear to a minimum and to dissipate the heat generated. In plain bearings, lubrication is achieved by maintaining a film of lubricant between the face either as a result of their relative motion or by pumping it through channels into the interface. Some small bearings, made of powdered copper or bronze impregnated with oil or graphite, are self-lubricating, while some low friction plastics (e.g.polytetrafluoroethylene) do not require lubrication. Bearing surfaces or shells are usually made of softer materials (e.g. brass or white metal) than is the shaft. In *ball and roller bearings* friction between the faces is reduced by replacing the sliding action of a plain bearing by a rolling action. The balls or rollers are usually made from chromium (0.5-2.0%) steels.

Beats. If two waves of nearly equal frequencies are superposed in a medium, the intensity of the resulting sound alternately increases and decreases. The waxing and waning of sound is termed as a beat. The frequency of the resulting signal (the beat frequency) is equal to the difference in frequencies of the two signals, i.e., $f_1 - f_2$. If two waves of equal amplitude (a) produce beats, the resulting amplitude (A) is given by:

$$A = 2a \cos [2\pi (f_1 - f_2)] t - \alpha/2$$

where, α is the phase angle between the original signals.

Bettie-Bridgeman Equation of State. This is an equation with five adjustable constants which can be made to represent the behaviour of a real gas above the triple point:

$$p = \frac{RT(1-\varepsilon)}{V^2} \cdot (V + B) - \frac{A}{V^2}$$

where p is the pressure of the gas, V is the volume, and

$$A = A_0\left(1 - \frac{a}{V}\right);\ B = B_0\left(1 - \frac{b}{V}\right);\ \varepsilon = \frac{c}{VT^3}$$

Beaufort Scale. A scale used in meteorology in which successive values of wind velocities are assigned numbers ranging from 0 (calm) to 12 (hurricane) to indicate wind forces. Numbers 13-17 are often added to indicate specific hurricane speeds.

Becke Line. When two adjacent substances, for instance a crystal immersed in a liquid ,are viewed under a microscope, then, if their refractive indices are close in magnitude, a narrow but bright line of light appears at the junction. The line is called the Becke line and arises from the combination of reflected and refracted light rays. If it is the crystal which has the higher refractive index, then as the objective of the microscope is raised, the Becke line appears to move into the crystal.

Becquerel. (Bq). The SI unit of activity (radioactivity) equal to the number of atoms of a radioactive substance that disintegrate in one second. Named after Antoine Henri *Becquerel.*

Becquerel Effect. If two similar electrodes are immersed in an electrolyte, a current flows when one of the elctrodes is illuminated.

Becquerel Rays. The radiation discovered by H. Becquerel (1896) to be emitted by uranium. Although uranium emits alpha, beta and gamma rays, they were beta rays which Becquerel detected using a photographic plate.

Beer-Lambert Law. The proportion of light absorbed by a dilute liquid medium chages exponentially with the product of the path length of the light in the medium, the molar concentration, and the molecular extinction coefficient.

Beer's Law. If I is the intensity of the light transmitted through a solution at a given wavelength and I_0 is the intensity of the incident light, then, according to Beer.

$$I = I_0\, 10^{-\mu cd}$$

where c is the concentration of solution in gram-molecules per litre and d is the thickness of solution through which the light is transmitted, μ is known as the molar extinction coefficient of the solute for the particular wavelength. The Law is sometimes used in the form

$$I = I_0 \exp(-acd)$$

where a is the molar absorption coefficient.

Bernoulli Effect (Bernoulli's Theorem). Bernoulli's equation gives a relationship between the pressure at a point in a steadily flowing liquid and the velocity of the liquid at that point. It is a consequence of the principle of conservation of energy and is,

$$\frac{p}{dq} + \frac{V^2}{2g} + h = \text{constant}$$

where, p is the pressure, V is the velocity of the liquid and h is the height of the point under consideration from a certain reference. The term $\frac{p}{dq}$ is called the pressure head, $\frac{V^2}{2g}$ velocity head and h the gravitational head. Thus, in the regions where the pressure is less, the liquid will flow through with large velocity and vice versa.

Bernuoulli's Principle. The principle of conservation of energy applied to fluid flow. If the effects of friction are neglected the total enegry of the flow at any point in a pipe is equal to the sum of the kinetic energy due to the flow velocity, the gravitaional potential energy due to height, and the energy of pressure in the fluid itself. Bernoulli's theorem states that the sum of these three components is constant throught a flow system. Named after Daiel *Bernoulli.*

Berthelot Equation. The Berthelot equation is a modified gas equation for real gases of the form

$$pV = RT\left\{1 + \frac{9}{128} \cdot \frac{pT_c}{p_c T}\left(1 - 6\frac{T_c^2}{T^2}\right)\right\}$$

where p_c and T_c are the critical pressure and critical temperature of the gas. The equation holds best at low pressures and is applicable only over a limited range.

Beta Current Gain Factor. Symboll: β The short-circuit current amplification factor of a transistor with common-emitter connection given by:

$$\beta = \left(\frac{\partial ic}{\partial iB}\right)_{VCE\ constant}$$

where *ic* is the collector current and i_B is the base current. β is always greater than unity and can takes values up to 500.

Betatron. A type of particle *accelerator* used for producing very high energy electrons. The electrons are accelerated round a circular path in an evacuated torus-shaped chamber, by means of a large pulsed magnetic field. Electron energies up to 300 MeV have been produced.

Betti's Reciprocal Theorem. If an elastic body is subjected to two system of body and surface forces, then the work that would be done by the first system of forces in acting through the displacements due to the second system of forces is equal to the work that would be done by the second system of forces acting through the displacements due to the first system of forces.

Bevatron . An accelerator that is used to accelerate protons and other particles to very high energies.

Big-bang or Superdense Theory. A theory in cosmology proposing that at sometime nearly 10×10^9 years ago all the matter of the universe was packed into a superdens small agglomeration, subsequently being hurled in all directions at enormous speeds by a cataclysmic explosion.

Binding Energy of Nucleus. Means the energy which is equivalent of the difference between the mass of the nucleus and the sum of the masses of its constituent nucleons. When this much of energy is supplied to the nucleus, it can undergo disintegration to yield the constituent nucleons far away from each other. The binding energy per nucleon is least for very light and very heavy nuclei and nearly constant (about 8 Mev/nucleon) for medium nuclei. So the fission of heavy nuclei to produce lighter nuclei and the fusion of very light nuclei to form heavier ones, liberate large amount of energy.

Bingham Fluid. A substance which remains rigid under a shear stress until the magnitude of the stress exceeds the yield stress, whereupon the substance flows like a *Newtonian Liquid.*

Binoculars. A portable optical instrument used for magnifying distant objects. It consists of two telescopes fixed side by side, one for each eye, inside which there are a number of lenses for magnifying and focusing the image and usually prisms for altering the direction of the light and thus increasing the effective length of the telescope.

Biophysics. The scientific discipline concerned with the explanation of biological phenomena in terms of the laws of physics. Biophysics emerged in the 1940s with the work of such scientists as Max *Perutz* and John *Kendrew,* who used *X-ray* diffraction to determine the structure of biological molecules; this was followed by Maurice *Wilkins'* work on DNA. More recent topics include the nature of the nervous impulse, the properties of biological membranes, the mechanism of muscle contraction, and the operation of sense organs.

Biot (Bi). A c.g.s. unit of current. 1 biot is the constant current which, if maintaned in two parallel rectilinear conductors of infinite length and neglible cross-section and with separation of 1 centimetre in a vacuum, would produce a force between the conductors equal to 2 dynes per centimetre of length. 1 biot = 10 amperes.

Biot-Fourier Equation. An equation for the conduction of heat through a solid:

$$\frac{\partial Q}{\partial t} = \frac{\lambda}{C_\rho} \Delta^2 T$$

where $\partial Q / \partial t$ is the rate of flow of heat, λ the thermal conductivity and C the specific heat of the solid of the density ρ.

Biot's Law. The rotation of the plane of polarization of light propagated through an optically active medium is proportional to the length of the path, to the concentration (if the medium is a solution) and approximately inversely proportional to the square of the wavelength of the light.

Biot-Savart Law. This law states that the magnetic field $d\vec{B}$ due to a small current carrying element of length dl through which a current I is flowing at a distance r from the element is given as follows:

$$d\vec{B} = \frac{\mu_0}{4\pi} I \frac{\vec{dl} \times \vec{r}}{r^3}$$

where, μ_0 represents a constant known as the permeability of vacuum, its value is $4\pi \times 10^{-7}$ weber- A-M. The law can be used for calculating the magnetic field due to currents flowing in various geometrical shaped wires.

Bire Fringence. (Also known as double refraction). When light is allowed to pass through crystals of certain materials like calcite, it splits into two plane-polarised rays, known as ordinary ray and extra-ordinary ray, plane-polarised in two mutually perpendicular planes. This phenomenon is known as fractions or bire fringence and the crystals which produce this effect are known as birefringent crystals.

Black Body. A theoretical body that absorbs all the electromagnetic radiation falling upon it. When heated it emits radiation (black-body radiation) having a coutinuous distribution of wavelengths with a maximum at a particular wavelength, which depends only on the temperature of the body.

Black Body Radiation. Refers to the radiation emitted by a black body. It, generally covers the entire wavelength range, the exact distribution of power depends on the temperature of the body.

Black Body Temperature. The temperature of a body as measured by a (radiation) pyrometer. It is generally appreciably less than the true temperature, T, of the body. For a temperature T_0 observed by a total radiation pyrometer, $T^4_0 = \varepsilon T^4$ where ε is the emissivity of the source.

Black Hole. A celestial "object" that has undergone such total *gravitational* collapse that no light can escape from it: its *escape* velocity exceeds the speed of light. Once a collapsing object's radius has shrunk below a critical value (the Schwarzschild radius) it becomes a black hole; for a star, this radius is about 10 km or less. The surface having this radius is called the event horizon of the black hole. The object will continue to contract until compressed to an infinite density at a single central point—a singularity. A black hole is thus a region of greatly distorted space (and time) the size of which increases with the mass of the contracting material.

No black hole has as yet been unambiguoulsly detected. The collapsed core remaining from the *supernova* explosions of massive stars are, however, promising candidates, especially if they are components of a *binary* star and thus more easy to detect. The X-ray binary Cygnus X-1 has a probable black-hole component. Black holes of immense mass (10^6 to 10^9 solar masses) are now thought to exist at the centres of certain galaxies, possibly including our own, and to be the powerful sources of energy in *active* galaxies such as *quasars*.

Blindness. Partial or complete loss of sight (degrees of visual impairment less severe than total loss of sight are classified as blindness for administrative purposes). Sudden blindness may be caused by direct injury to the eye or to the part of the brain that receives the visual signals. Blindness that develops gradually is caused by a wide variety of diseases, including *glaucoma, cataracts,* diabetes mellitus, and tumours (especially of the pituitary gland), that compress the optic nerve. There are various aids available for the visually handicapped, including books in *Braille* or on tape or records and specially trained *guide* dogs. In many countries the blind are entitled to special education and financial benefits.

Blind Spot. The region of the retina where the optic nerve leaves the eye.

Bloch's Theorem of Superconductivity. Bloch proved that the lowest state of a quantum-mechanical system in the absence of a magnetic field can carry no current.

Bloch Wall. A ferromagnetic matêrial in zero magnetic field will be in equilibrium if it has minimum potential energy. This cannot be so if the crystal has a single domain of one direction of magnetization. So the crystal consists of domains which are magnetized in different directions. Bloch walls are the transition layers between domains which occur in bulk samples. The magnetic spin orientation of the atoms gradually changes within the layer, with the magnetic spin vector rotating about an axis perpendicular to the boundary. This arrangement is energetically more favourable than one in which the axis of rotation is in the plane of the boundary. For the latter case, free poles would produced over the whole boundary wall and this would correspond to a larger potential energy. However, for thin film samples the second case is energetically more favourable and the wall then formed is called a *Neel Wall.*

Blondel. A metric unit of luminance. One blondel is the luminance of a uniform diffuser emitting 1 lumen per square meter.

$$1 \text{ blondel} = \frac{1}{\pi} \text{candela metre}^{-1}.$$

Blondel-Rey Law. A light source, flashing at low frequency (less than 5 hertz) and having brilliance B_0 during the flash, has an apparent brilliance B given by

$$B = \frac{B_0\, t}{\alpha + 1}$$

where *t* is the time of duration of the flash and α is a constant (equal to approximately 0.2 s when B is near the threshold for white light). *See also* Talbot's Law.

Bode's Law. Bode stated that, if the distance of Mercury from the Sun is taken as 4, the corresponding distances of the other planets would be given, in the same units , as : Venus 4+3, the Earth 4 +6, Mars 4 + 12, Asteroids 4 +24, Jupiter 4 +48, Saturn 4+96 and Uranus 4 + 192. Neptune however, which should lie according to Bode's law at a distance of 388 from the Sun, was found to be in fact 300 units away. Sometimes called the *Titius-Bode Law.*

Bohr-Breit-Wigner Theory. A nuclear reaction occurs in two stages. In the first, the bombarding particle is absorbed by the target nucleus to form a high energy unstable compound nucleus. In the second, this nucleus emits radiation to leave a residual lower energy nucleus known as a recoil nucleus.

Bohar Magneton (μ_B)

A unit of magnetic moment.

$$\mu_B = \frac{eh}{4\mu m} = 9.2741 \times 10^{-24} \text{ joule tesla}^{-1}$$

In c.g.s. units, μ_B equals 9.2741×10^{-21} erg gauss^{-1}.

Bohr's Correspondence Principle. The behaviour of the electrons in an atomic system must approach more and more that predicted by classical physics the higher the quantum number of the orbit.

Bohr-Van Leeuwen Theorem. These workers have shown that, in any dynamical system to which classical statistical mechanics can be applied, the paramagnetic and diamagnetic susceptibilities cancel, so that, classically, for thermal equilibrium, one always has zero susceptibility.

Bohr-Wheeler Theory of Fission. A theory of fission based on a liquid-drop model of the nucelus in which there are coulombic and surface tension forces.

Boiling Point. The constant temperature at which a liquid boils and at which the saturated vapour pressure of the liquid is equal to the external pressure.

Boltzman Constant. (*k*): A constant $k = \frac{R}{NA}$ where R is the gas constant and N_A is the Avogadros number. Its value is 1.38 J/K.

Boltzmann Equation. The entropy of a system of particles is proportional to the logarithm of the probability of its macroscopic state. The constant of proportionality is known as *Boltzmann's Constant.*

Boltzmann H Theorem. The *Boltzmann H function* is defined as

$$H = \int\int \ln f \, dv \, dr$$

where integration is over all position and momentum space, f is a distribution function and v and r are velocity and position vectors respectively. Then, the Boltzmann H theorem states that H is a decreasing function with time unless F is Maxwellian in which case H remains constant. This is analogous to the thermodynamic statement that the entropy of a system tends to a maximum, and, except for an additive constant, the H function equals the negative of entropy divided by Boltzmann's constant.

Boltzmann's Law of Radiation. (Stefan-Boltzmann Law). Stefan induced an experimental law for the intensity of the total radiation from a black-body at various temepratures. This law was subsequently derived theoretically by Boltzmann. The law states that the complete emission W of a black-body is proportional to the fourth power of the absolute temperature.

$$W = a T^4$$

where a is a constant which includes *Stefan's Constant.*

Boltzmann, Ludwig Eduard (1844-1906). Austrian physicist, who developed statistical mechanics with J.C. *Maxwell* and J.W. *Gibbs,* notably the Maxwell-Boltzmann statistics of particle systems obeying classical laws. He also linked thermodynamics with molecular physics by showing that increasing entropy is related to increasing disorder among particles. Working with his teacher Josef Stefan (1835-93), he showed that Stefan's law could be derived thermodynamically and it is now usually known as the Stefan-Boltzmann law.

Boltzmann's Ratio. Where particles can have two energy levels not widely separated, the distribution of the particles over these levels is controlled by Boltzmann's ratio

$$\frac{n_2}{n_1} = \exp\left(\frac{-\Delta E}{k_B T}\right)$$

where ΔE is the energy difference between the levesl and n^2 particles are in the upper level and n^1 in the lower level.

Boltzmann's Transport Equation. An equation expressing the variation $(f—f_0)$ of a distribution function for charged carriers from its equilibrium value f_0 in the presence of a thermal gradient and fields such as electric and magnetic fields. If v is the velocity of the carriers, p their momentum and *dp/dt* the force on them at time *t*, and τp is a relaxation time for the collision processes, then, for elastic scattering of the carriers and small perturbations, it can be written in the form

$$\frac{\partial f}{\partial t} + v\nabla f + \frac{dP}{dt}\nabla_P f = \frac{f - f_0}{\tau(p)}$$

For steady-state process $\partial f / \partial t = 0$

Booster. (1) A generator connected in series with an electric circuit to enable the voltage acting in the circuit to be increased or decreased. (2) In broadcasting, a repeater station that receives the signal transmitted form a main station, amplifies it, and then retransmits it, sometimes with a change infrequency.

Bootstrap Theory. The theory in which no elementary particle is regarded as being more fundamental than any other. Each particle exists by virtue of the existence of all the others. Its name derives from the phrase "to pull oneself up by the bootstraps." The theory avoids the problem of a series of classes of particles, each more fundamental than the last.

Boson. A class of elementary particles with integral spin. Bosons always obey *Bose-Einstein* statistics and include all the mesons and the *Photon.*

Bosanquet's Law. The resistance of an electric circuit is defined as the ratio of the electromotive force to the current *(Ohm's Law)*. In a magnetic circuit the ratio of the magnetomotive force F_m to the magnetic flux ϕ is known as the relutance R. Bosanquet's Law states that

$$\phi = \frac{F_m}{R} \quad \text{in analogy to Ohm's Law.}$$

Bourdon Gauge. A pressure gauge consisting of a C-shaped closed tube that tends to straighten as the pressure of fluid inside it rises; this movement is transferred to a pointer or a digital display. Invented by E. Bourdon (1808-84).

Bow's Notation. A metod of notation of forces action at a point, wherein the spaces between forces are lettered or numbered, any force being defined by the letters or numbers on each side of it.

Boyle, Robert (1627-91). Irish physicist and chemist. Boyle showed that air possesses weight, it is necessary for sound propagation, and its pressure affects the boiling point of water. His chemical work, summarized in *The Skeptical Chymist* (1661), distinguished between elements, compounds, and mixtures and dismissed the Aristotelian concept of the four elements. His work on gases, with Robert *Hooke,* led to the law that bears his name, Boyle published his law in 1663, but in France it is named after Edme Marriotte (1620-84), who did not publish it until 1676.

Boyle's Law (Marriotte's Law). At a constant temperature the volume of a definite mass of gas is inversely proportional to the presure, i.e.

$$pV = \text{constant}.$$

Real gases deviate appreciably in behaviour from this equation, as might be expected as the law may be deduced from kinetic theory, which considers the molecules of the gas as point masses having no attraction for one another.

Boy's Radiomicrometer. A thermocouple and moving coil galvanometer combined to make a sensitive detector of radiant energy.

Bragg-Gray Principle. The energy absorbed per unit mass $(dE/dm)_b$ for a particular medium b is related to the ionization which occurs in a small gas-filled cavity in that medium by

$$(dE/dm)_b = P_b W_g J_g$$

where P_b is the mass stopping power of the medium relative to the gas g, W_g is the average energy required to produce an ion-pair in a gas and J_g is the number of ion pairs produced per unit mass of gas in the cavity (determined experimentally). The principle has application to the measurement of ionizing radiation.

Bragg-Kleenan Rule. The atomic stopping power of a material to charged particles is proportional to the square root of the material's atomic weight. Atomic stopping power is additive and thus can be obtained for complex molecules.

Bragg-Pierce Law. The atomic absorption coefficient of an element for X-rays is given by

$$\mu_a = CZ^4\lambda^3 \text{ (sometimes quoted } \mu_a = C'Z^5\lambda^{7/2})$$

where Z is the atomic number, λ is the wavelength of the X-rays and the constant C (or C') varies with λ.

Bragg's Law. A law stating that, if two parallel X-rays of wavelength λ are reflected by adjacent planes, distance d apart, in a crystal lattice and the rays then constructively interfere, then $2d \sin\theta = n\lambda$, where n is an integer and θ the angle between the X-rays and the planes. θ is called the Bragg angle. If n is half-integral, then destructive *interference* is observed. Named after Sir William and Sir Lawrence *Bragg*.

Bravais' Law. Those crystal forms which tend to occur most frequently are those with faces parallel to planes of smallest recticular area. (Recticular area is inversely proportional to the interplanar spacing).

Breit-Wigner Formula. If a nucleus is excited to a well-defined single excitation level byan incoming particle *n*, for instance a neutron, with the emission of a particle *p,* then the reaction cross-section is given by

$$\frac{\pi g}{k^2} \frac{\Gamma_n \Gamma_p}{(E - E_r)^2 + (\Gamma/2)^2}$$

where $$g = \frac{2J + 1}{(2I + 1)(2S + 1)}.$$

J is the spin of the compound nucleus, I the spin of the target nucleus and *s* the spin of the incoming neutron *n*. Γn and Γp are the neutron and particle energy widths and Γ is the total energy width of the compound nucleus (proportional to the disintegration constant). E_r is the energy at the peak of the resonance, E the total energy (of the target and incoming neutron) and *k* the wave-vector associated with the incident neutorn.

Brewster (B). The unit used to measure the stress-optical constant when a material shows bire fringence under an applied stress. A brewster is the number of angstroms per millimetre path by which one component of the light is retarded relative to the other if a stress of 1 bar (atmosphere) is applied in a direction perpendicular to the path of the light.

$$1\ B = 10^{-12}\ Pa^{-1}.$$

Brewster's Fringes. Intreference fringes observed when light from an extended source passes through two etalons (pairs of parallel semi-silvered glass plates with an air gap between) which are inclined at a small angle. There is a path difference established between light waves reflected in the two etalons, which must be

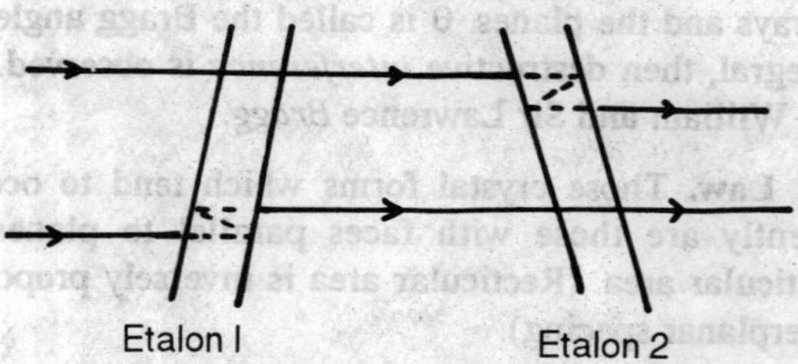

of equal thickness or whose ratio of thickness is an integer. Usually, transmission fringes are observed, but the reflection fringes in a Jamin interferometer are a particular case of Brewster's fringes.

Brewster's Law. When a beam of light is reflected from a surface, polarization may occur. As the angle of incidence is increased from normal to grazing the polarization increases, passes through a maximum and then decreases. The angle of maximum polarization is known as the angle of polarization or *Brewster angle* and varies with the nature of the substance. Brewster discovered that the refractive index was equal to the tangent of the angle of polarization. This law leads to the result that the sum of the angle of incidence and angle of refraction at maximum polarization is equal to a right angle.

Brillouin Function. In a paramagnetic substance the magnetization can be shown to be

$$M=Ng\,J_{\mu B}\,Bj(x)$$

where N is the number of atoms per unit volume each with a total angular momentum quantum number J and a spectroscopic splitting factor *g* (*Lande Splitting Factor*), μ_B is the *Bohr Magneton,* and

$$x = \frac{gJ_{\mu B}H}{kT} .$$

B_J is the Brillouin function

$$Bj = \frac{2J+1}{2J}\coth\frac{(2J+1)x}{2J} - \frac{1}{2J}\coth\frac{x}{2J} .$$

Brillouin Scattering. Scattering of electromagnetic waves in solids and liquids when, as a result of the scattering process, an acoustic phonon is emitted or absorbed. (An acoustic phonon is a lattice vibration in which atoms in a unit cell vibrate in the same direction. The dispersion curve for the vibrations is of the same form as for sound waves). Brillouin scattering is analogous to *Raman Scattering*.

Brillouin Zone. If the reciprocal lattic of a crystal is obtained, but with periodicity of the lattice $2\pi/d_{hkj}$ instead of $1/d_{hkl,}$ and the space be divided into identical cells in the same way as the real lattice is divided to obtain the *Wigner-Seitz Cell,* then the cells so obtained are called Brillouin zones. These zones are very important in the theory of energy bands in crystals because, by *Bragg's Law,* the zones may be said to define a momentum or *k* space in which *k* is the wave vector for the electrons. If at the reciprocal lattice points there is strong reflection of electrons or X-rays (this will be so, unless the so-called structure factor used in X-ray analysis is zero), then there will be an energy gap at each face of the Brillouin zone.

Brinell's Hardness Test. The hardness of a material is measured by the area of indentation produced by a standard steel ball of 10 mm diameter under prescribed conditions of loading-usually 3000 kg *f* (29.4 kN) static load for steel and 500 kg f (4.9 kN) for softer metals. The *Brinell number* is given by

$$\frac{P}{\pi t D} = \frac{P}{\pi D\{D/2 - \sqrt{(D^2/4 - d^2/4)}\}}$$

where *p* is loading in kgf, *t* is depth of indentation, D diameter of the ball and *d* diameter of the indentation, all in millimeters.

British Thermal Unit (btu). The quantity of heat required to raise the temperature of 1 lb of water by 1° F. btu= 1055.06 J or 251.997 calories.

Broca Galvanometer. An instrument with an astatic system formed by two magnet capable of movement, each with a consequent pole in the middle.

The arrangement combines a large magnetic moment with a small moment of inertia, and gives a high sensitivity with a reasonably small period.

Bronson Resistance. An ionization chamber that encloses a constant source of ionization, generally a layer of uranium oxide. If the potential difference between the electrodes is small compared with that required for saturation, the ionization current is approximately proportional to the applied p.d., and the system acts as a high resistance.

Brooks-Herring Formula. An expression for the mobility of electrons in a medium containing a number of point imperfections or impurities. It takes the same form as the *Conwell-Weisskopf Equation* but replace 1/ln(1+ x) by F ($3K_B$T) where F(E) is a Fermi-Dirac function.

Bubble Chamber. An instrument that makes visible the tracks of ionizing particles, used for observing particle decays and interactions. It contains a liquid, usually hydrogen, helium, or deuterium, under pressure and at a temperature slightly above its normal boiling point. The particle induces boiling along its path and the bubbles are photographed to record the particle's track and those of any charged decay or reaction products.

Budde Effect. The expansion observed on exposing halogen vapours to light caused by the rise in temperature resulting from heat evolved in the recombination of atoms.

Buffer. An isolating circuit that minimizes reaction between two circuits. It has a high input impedence and low output impedance.

Bunsen Ice Calorimeter. A calorimeter to measure the specific heat capacity of a solid or liquid when only a small quantity is available.

Buoyancy. When a body is wholly or partly immersed in a fluid, it experiences an upward force equal to the weight of the fluid. This force known as the buoyance, acts through the centre of gravity of that fluid, this point is known as the *centre of buoyancy* of the body.

Burger-Dorgelo-Orstein Rule. For small multiplet splitting in atomic spectra when *Russell-Saunders Coupling* is applicable, the sum of the intensities of all the lines of a multiplet which belong to the same initial or final state is proportional to the satistical weight of 2J+1 of the initial or final state respectively, where J is the quantum number for the total angular momentum of the electrons.

Burgers' Circuit. A method of determining whether an imperfection is contained in a crystal lattice. Consider the perfect two-dimensional structure in diagram (a). The circuit marked with the arrows consists of ten moves to go from A back; e.g. 1→2→3 and so on. In diagram (b), a dislocation is contained in the area of Burgers' circuit and it now takes more than ten moves to go from A back to A

Burstein Effect. The variation in width of the optical energy gap of certain semiconductors with the amount of doping. Usually, the optical gap is between the highest energy state in the valence band and the lowest state in the conduction band; i.e. the smallest energy difference required for excitation of electrons form the valence to the conduction band. In some semiconductors having a very large carrier density, the first unoccupied states lie well within one of the bands and this position is a function of doping.

C

Caesium Clock. An *atomic* clock that depends on the energy difference between two states of the caesium nucleus in a magnetic field. A non-uniform magnetic field is used to split a beam of caesium atoms into two components. Nuclei in the lower energy state are irradiated by radio-frequency radiation at the difference frequency between the two states, so that some are excited to the higher state. By reanalysing the mixture, the radio-frequency oscillator can be locked to the difference frequency of 9 192 631 770 hertz with an accuracy of one part in 10^{13}. This extremely accurate clock is now used in the definition of the second.

Cailletet and Mathias Law. The mean of the densities of a liquid and its saturated vapour at the same temperature is a linear function of temperature. The densities of the liquid and of saturated vapour in equilibrium with it are known as orthobaric densities and, if ρ_t is the arithmetical mean at any temperature t, then

$$\rho_t = \rho_0 + \alpha t$$

where ρ_0 and α are constant for a given substance.

Callender and Barne's Apparatus. Apparatus used to measure the specific heat of a liquid, or to find the value of Joule's mechanical equivalent of heat. It consists of a long tube having a heating wire along its axis. The liquid is allowed to enter the tube at one end leave at the other—while a current I is passed through for the heating wire. If represents the mass of the liquid flowing per second and v the potential difference across the wire, $VI = ms(\theta_1 - \theta_2) \times J$ where θ_2, θ_1 are the temperatures of outgoing and incoming water. The value of S or J can be calculated.

Callendar's Equation. An equation representing the behaviour of superheated steam below the critical point and away from the liquid vapour region. It is:

$$V - b = \frac{RT}{p} - \frac{a}{T^n}$$

where $n = 10/3$ and a and b are constants.

Calorie. A unit of heat now being replaced by the Joule. Formerly defined as the amount of heat required to raise the temperature of 1 grm of water through 1^0 C, the calorie is now defined as 4.1868 joules. The kilocalorie or Calorie (with a capital "C") is equal to 1000 calories and was formerly used to express the energy value of foods.

Calorific Value. The amount of heat liberated on complete combustion of unit mass of a fuel. The determination is carried out in a bomb calorimeter and the value is usually expressed in J kg^{-1}.

Calorimeter. Any vessel used to carry out thermal measurements quantitatively.

Calorimetry. Refers to the study of the measurement of quantities of heat.

Calutron. An electromagnetic separator of isotopes, used to separate the fissile radioisotope uranium-235 from uranium-238.

Calzecchi-Onesti Effect. The conductivity of loosely aggregated metallic powders changes under the action of applied electromotive forces.

Camera, Photographic. A device for producing a photographic image. Basically, a camera consists of a light-tight box containing a lens and light-sensitive film or plate. The light image coming through the lens is brought into focus on the film by adjusting the distance between the film and the lens. A picture is taken by opening the shutter over the lens for a certain period to expose the film. The exposure time is determined by the shutter speed. The diameter of the opening (aperture) in front of the lens is measured by its *f-number*. Shutter speed and lens aperture determine the light availabe to record the image. In the simplest cameras they are fixed, but sophisticated cameras have variable settings. Sometimes the aperture or the shutter speed can be controlled automatically by an *exposure meter*.

Camera, Television. A camera for the instantaneous transmission of *television* pictures. The scene to be televised is focused onto a screen in the electronic camera tube. This optical image is scanned in horizontal lines, fast enough to appear as a continuous moving picture to the human eye; in most cameras, the image is scanned 25 to 30 times per second. The intensity of light at each point of the image is converted to an electrical signal, which is amplified before being transmitted together with the sound and synchronization signals.

Candela (cd) .The SI unit of luminous intensity defined since 1979 as the intensity in a given direction of a source emitting monochromatically at 540×10^{12} hertz with a radiant intensity in that direction of 1/683 watt per steradian.

Candle Power. It is the luminous intensity of a light source expressed in terms of the candela.

Capacitance. The ability of an electrical component to store charge. It is meausred in *farads* and defined as the ratio of the stored charge in coulombs to the voltage drop across the component. Capacitance is one of the factors that control the frequency response of circuits and components to alternating currents.

Capacitor. An electrical component (formerly called a condenser), with an appreciable *capacitance*. It consists of two conductor or semiconductor plates separated by a *dielectric*. The value of the capacitance is a function of the geometry and the electrical properties of the dielectric and often also of the operating voltage and the frequency. Variable capacitors are used for tuning electronic circuits.

Capacitors in Series and Parallel. The capacitors are said to be connected in series if the outer coating of one is connected to the inner coating of the next and so on. The resultant capacity of the combination (C) is given as follows:

$$\frac{1}{C} = \frac{1}{C_1} + \frac{1}{C_2} + \frac{1}{C_3}.$$

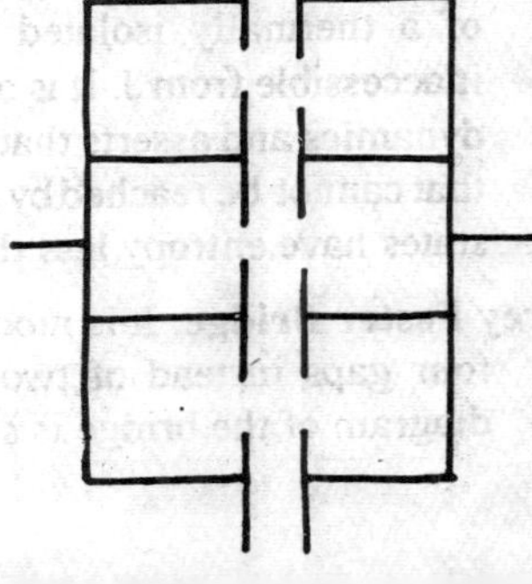

They are said to be connected in parallel if the outer plate of all of them are connected at one point and the inner plates of all at other point. The Resultant capacitance of the combination is given by

$$C = C_1 + C_2 + C_3 + \ldots\ldots\ldots\ldots$$

Capillarity. A term to describe the rise or fall of liquids in vertical capillary tubes. The elevation of a liquid in a capillary tube is given by Juvin's rule.

Capillary Electrometer. An electrolytic cell in which one electrode is a pool of mercury A, while the other is the meniscus B of a thread of mercury in a capillary tube, CD. If a potential difference is applied to the electrodes, a small charging current flows through the cell and produces polarization at the meniscus B. The electric field due to this polarization changes the surface tension of the mercury and the meniscus moves to a new position of equilibrium. The movement of the mercury is not strictly proportional to the applied p.d., and the electrometer is most conveniently used as a null instrument.

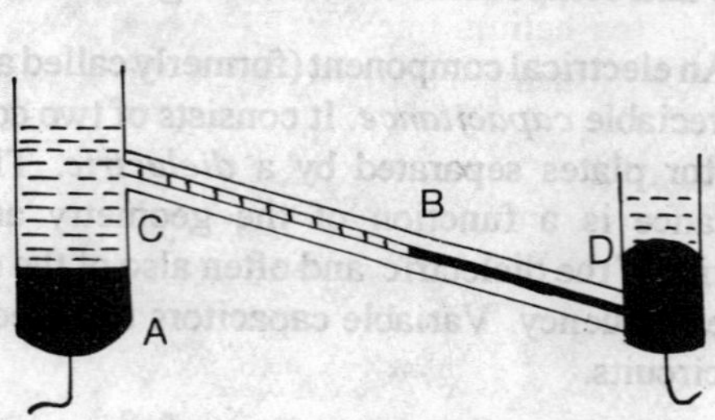

Capture. Refers to a process by which an atom, ion, molecule, or nucleus acquires an aditional particle.

Caratheodory's Principle. In the neighbourbood of an arbitrary state J of a thermally isolated system there are states J' which are inaccessible from J. It is a statement of the second law of thermodynamics and asserts that there are states as close as we wish to J that cannot be reached by an adiabatic process. These inaccessible states have entropy less than J.

Carey Foster Bridge. It is modified form of a slide wire bridge. It has four gaps instead of two in the slide wire bridge. The circuit diagram of the bridge is shown in the figure below.

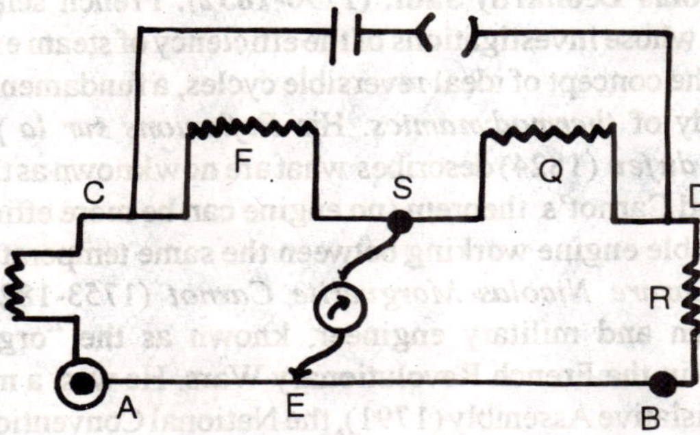

P,Q are two approximately equal resistances. R is a known resistance and X is the unknown resistance whose value is to be determmed. It is very accurate for determination of low resistances.

Carbon Cycle (Physics). A cycle of thermonuclear reactions in which a nucleus of carbon-12 acts as a catalyst in converting four hydrgen nuclei into a helium nucleus. The cycle produces energy and is believed to be a major source of energy in some stars. The reactions are:

$$ {}^{1}_{1}H + {}^{12}_{6}C \rightarrow {}^{13}_{7}N \rightarrow {}^{12}_{6}C + {}^{0}_{1}e $$

$$ {}^{1}_{1}H + {}^{13}_{6}C \rightarrow {}^{14}_{7}N $$

$$ {}^{1}_{1}H + {}^{14}_{7}N \rightarrow {}^{15}_{8}O \rightarrow {}^{15}_{7}N + {}^{0}_{1}e $$

$$ {}^{1}_{1}H + {}^{15}_{7}N \rightarrow {}^{12}_{6}C + {}^{4}_{2}He. $$

Carbon Dating (Radiocarbon Dating). A method used of determining the age of materials, usually archaeological, that contain matter of living origin. It consists in determining the ratio of ${}_{6}C^{12}$ to the ${}_{6}C^{14}$ (a radioactive isotope of carbon with a half life of 5730 years). The ratio enables us determine the time when the death of the sample has occured.

Carnot-Clausium Equation. For a reversible closed system the total change in the entropy of the system, *dq*/T, is zero; *dq* is the quantity of heat absorbed by the system during an infinitesimal reversible change of state and T is the absolute temperature of the system during this change.

Carnot, (Nicolas Leonard) Sadi. (1796-1832). French scientist and solider, whose investigations of the efficiency of steam engines led him to the concept of ideal reversible cycles, a fundamental idea in the study of *thermodynamics.* His *Reflexions sur la puissance motrice du feu* (1824) describes what are nowknown as the *Carnot* cycle and Carnot's theorem (no engine can be more efficient than a reversible engine working between the same temperatures). His father *Lazare Nicolas Marguerite Carnot* (1753-1823) was a statesman and military engineer, known as the "organizer of victory" in the French Revolutionary Wars. He was a member of the Legislative Assembly (1791), the National Convention (1792), and the Committee of Public Safety. Carnot was exiled in 1815 and settled in Magdeburg. His *De la defense de places fortes* (1810) became a classic work on fortifications.

Carnot Cycle. A reversible thermodynamic cycle of changes of *pressure* and temperature in the gas in an ideal heat engine. The gas is compressed adiabatically (at constant heat content), thus raising its temperature, say from T_1 to T_2. It is then expanded isothermally (at constant temperature). The gas is then expanded adiabatically, lowering its temperature from T_2 back to T_1, and finally compressed isothermally at T_1, thus completing the cycle. The efficiency of this cycle depends not on the nature of the gas but only on the temeperature range, i.e. it is equal to $(T_1—T_2)/T_1$, where T_1 and T_2 are absolute temperatures. Named after Sadi *Carnot.*

Cascade. Cascade means a number of capacitors connected such that the outer plate of the first capacitor is connected to the inner plate of the next, end so on. All the plates are insulated except the last. The capacitance of the compound capacitor is given by $1/C = 1/C_1 + 1/C_2 + ...$ The charges on all the capacitors are equal.

Casade Liquefier. A gas liquefier where in the critical temperature of the gas, below which it must be cooled before it can be liquefied, in steps. The steps use a series of gases having successively lower boiling points.

Cataract (Ophthalmology). Opacities in the lens of the eye resulting in blurred vision and caused by the deposition of small crystals of changes in the composition of the lens substance. The former condition increases with age; such cataracts are a common cause of blindness in the elderly. Certain diseases, such as poorly controlled diabetes mellitus, can also lead to cataracts. Cataracts are usually treated by surgical removal of the lens and the use of appropriate spectacles. In some cases a plastic lens may be implanted to replace the one removed.

Catenary. The curve obtained by suspending a string between two points. If the middle of the string, the lowest point, is at height *a*, above a reference level, then the height, *y*, at distance *x* along the string from the middle is given by: $y = 1/2a\ (e^{x/a} + e^{-x/a})$.

Cathetometer. Refers so a telescope mounted on a graduated vertical pillar along which it can move. Used for measuring displacements at a distance of a few feet.

Cathode-Ray Oscillosscope (CRO). An instrument that displays electrical quantities on the screen of a *cathode-ray* tube. It can be used to show the variation of a signal strength with time or with another electrical quantity. The CRO is used extensively in scientific work.

Cathode Rays. A stream of electrons emitted by a *cathode,* when a voltage is applied between a cathode and an *anode* either in an evacuated glass tube or one containing gas at low pressure. The electron beam can be focused onto a fluorscent screen to produce a visual display. This effect is used in the *cathode-ray* tube used in television receivers, radar screens and oscilloscopes.

Cathode-Ray Tube (CRT). A vacuum tube that converts electrical signals into visible form by projecting a beam of electrons onto a fluorescent screen. It is an essential component of the television receiver and the *cathode-ray* oscilloscope (CRO). The electron beam is produced by an electron gun, and deflected horizontally and vertically by an arrangement of plates and magnets, which move it back and forth across the screen and focus it by creating an *electromagnetic* field, the strength of which.varies according to input signals. In a television tube the beam intensity varies to form the light and dark regions of the picture.

Cauchy Dispersion Formula. A formula for the dispersion of light of the form: $n = A + (B/\lambda^2) + (C/\lambda^4)$, where n is the refractive index, λ the wavelength, and A, B and C are constants. Sometimes only the first two terms are necessary.

Celestial Mechanics. The study of the motions of celestial bodies subject to mutual gravitational interaction through the application of the laws of *gravitation* and of *mechanics*.

Celestial Sphere. The imaginary sphere, of immense size, at the centre of which lies the earth and on the inner surface of which can be projected the stars and other celestial bodies. The directions of these bodies, as seen from earth, are measured in terms of their angular distances from certain points and circles on the celestial sphere. These circles include the *ecliptic,* the observer's horizon, and the *celestial equator,* where the earth's equatorial plane meets the celestial sphere. The reference points include the equinoxes, the zenith and the *celestial poles,* where the earth's axis meets the celestial sphere. The earth's daily rotation causes an apparent and opposite rotation of the celestial sphere.

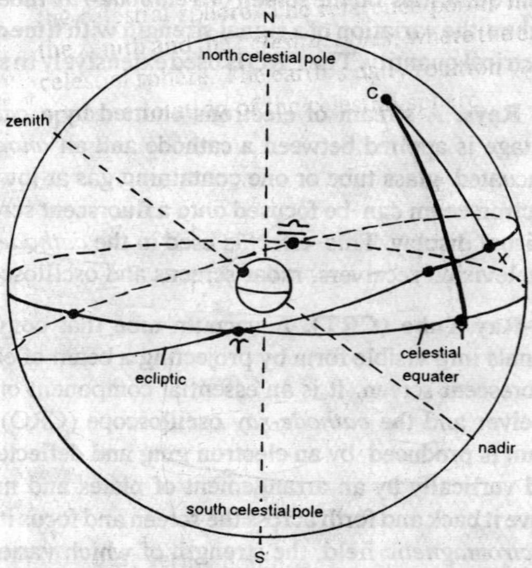

♈	first point of Aries; vernal equinox
♎	first point of Libra; autumnal equinox
C	celestial object

ϒY	right ascension of C (in hours anticlockwise from ϒ)
ϒX	celestial longitude of C (in degrees anticlockwise from ϒ)
YC	declination of C
XC	celestial latitude of C

CELESTIAL SPHERE

Cell. In an *electrolytic cell,* an e.m.f. is developed across the two electrodes by a chemical reaction.

Cell Constant. The constant obtained by dividing the area of the electrode in an electrolytic cell with the distance between the electodes.

Celsius Scale. A temperature scale devised by the Swedish astronomer Anders Celsius (1701-44), who originally designated zero as the boiling point of water and 100^0 as the freezing point. The scale was later reversed. Celsius temepratur, t, is defined by the equation $t = T - 273.15$, where T is the *thermodynamic* temperature. Each degree Celsius is equal to one *kelvin.* Until 1948 it was known as the centigrade scale.

Centre of Gravity. A fixed point through which the resultant gravitational force on a body always passes no matter what its orientation.

Centre of Mass. A point at which the mass of the body may be supposed to be concentrated.

Centrifugal Force. A force that acts radially outwards on a body moving in a circle of radius r with speed v. The magnitude of this force is given by mv^2/r where, m is the mass of the body.

Centrifugation Potentials. An electric field generated in a colloidal solution when centrifuged. The disperse phase moves away from the rotational axis retaining its charge. The smaller counter-ions are left behind. This gives rise to a p.d. between points in the solution at different distances from the axis. It can be used as a means of determining the zetapotential in colloidal solutions.

Centrifuge. A device for rotating mixtures of substances at high speed so that the heavier components can be separated from the lighter;

the heavier comonents experience a greater centrifugal force, thus enabling components of different densities (e.g. milk and cream) to separate. The ultracentrifuge is a high-speed device used in the determination of molecular weights and in biochemistry. Large centrifuges are used in physiological research and in training programmes to obtain a high *acceleration* of free fall (*g*).

Centripetal Force. A force that acts on a moving body causing it to move in a curved path. An object is constrained to move in a circle by a centripetal force directed towards the centre. It has an acceleration towards the centre of v^2/r, where v is the object's velocity and *r* the radius of the circle. The centripetal force is necessary to overcome the object's tendency to move in a straight line; it appears to be balanced by an equal *centrifugal force* directed radially outwards. However, this is a fictitious force. In the case of a satellite (mass *m*) orbiting the earth (mass M), the centripetal force is the gravitational force GmM/r^2, where G is the gravitational constant (*see* gravitation) and r is the height of the satellite above the centre of the earth. Thus $GmM/r^2 = mv^2/r$, which enables the height of the orbit to be calculated for a given value of *v*. It is sometimes said that the satellite stays in orbit when the centrifugal force balances the gravitational force, but is the centrifugal force is fictitious this causes confusion; the gravitational force is not balanced by the centrifugal force; it is the centripetal force.

Century. The ancient Roman unit of a hundred. In the Roman army centuries were the 60 subdivisions, composed of about a hundred infantrymen, of a legion and were commanded by centurions. Centuries were also political divisions of Roman citizens, each of which had a group vote in the assembly of centuries.

Centroid. The centre of mass of a uniform solid with the same shape as the volume.

Centrosymmetry. Having Symmetry with respect to a point. Crystals that are centrosymmetrical have their faces arranged in parallel pairs these pairs are alike or enantiomorphous in surface properties.

Cerenkov Radiation. Refers to the bluish light emitted by a beam of high-energy charged particles passing through a transparent medium at a speed, *v*, it is greater than the velocity of light *c*, in that medium. The radiation represents the excess energy resulting from the difference in velocity of the particle and the velocity of associated electric and magnetic fields which cannot exceed the velocity of light in the medium. Cerenkov radiation is analogous to the shock wave produced by a sonic boom.

C.G.S. System. A metric system of units based on the centimetre, gram and second. It includes dynes, ergs, and both electrostatic and electromagnetic units. It has been replaced for scientific purposes by *SI units.*

Chain. (1) A measure of length equal to 66 feet. (2) A measuring device used in surveying. A Gunter's chain is 66 feet long (consisting of 100 links), whereas an Engineer's chain is 100 feet long (also 100 links).

Chain Reaction. In nuclear fission, some heavy nuclei split into lighter nuclei by absorption of neutrons the resulting products are known as fission fragments. Each fission event also liberates several neutrons. The liberated neutrons produce fission in the remaining heavy nuclei and a rapidly multiplying sequence of fission events takes place. Such a sequence is known as a chain reaction. A chain reaction can be initiated ad controlled in a nuclear reactor. For any chain reaction to take place, it is essential that a certain minimum amount of the fissionable material called, the critical mass is initially present.

Chandrasekhar Limit. No white-dwarf star can be more massive than approximately 1.4 solar masses. The limit arises because the large the white-dwarf star, the closer it packs; the upper mass limit is set by how close the atoms can be compressed together without producing a neutron star, where the protons and electrons are forced to combine by the gravitational pressure.

Characteristic X-Radiation. The characteristic spectrum of an element may be obtained by making it the target in an X-ray tube.

Charge. Symbol: Q. A property of some elementary particles which causes them to exert forces on one another. Charge is the integral of electric current with time and is measured in coulombs. The electrons has a charge of $1.602\ 192 \times 10^{-19}$.

Charge Conjugation Partity. Syn. C-parity. Symbol: C.A. quantum number associated with such elementary particles (such as π^0 and η) that have zero charge, baryon number, and strangeness. It shows whether the wave function describing the particle is unchanged (C = + 1) or changes sign (C = — 1) when the particle is replaced by its antiparticle.

Charge Density. The charge per unit volume of a material, per unit area of a surface or per unit length of a line. They are respectively called volume, surface—or line charge density.

Chemical Hygrometor. A hygrometer in which water flowing from the aspirator D causes a known volume of air to be drawn through E, any moisture contained in the air is removed in the drying tubes A, B and C. The process is repeated with the tube R L, containing water, attached at E. Air passing through R L becomes saturated with water vapour. The ratio of the increase in weight of A and B before R L is atached and after gives the relative humidity of the atmosphere.

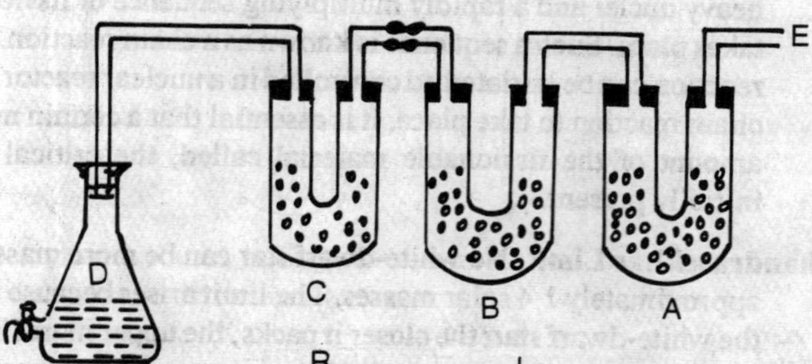

Chemical Shift. A change in the position of a spectrum peak that result from a small change in energy level due to a chemical effect.

Cherenkov Radiation. Radiation which arises when a charged particle traverses a medium at a velocity greater than the phase velocity of light in that medium. The particle sets up local polarization of the dielectric medium and, as the coherent growth and decay of a sequency of dipoles occurs along the path of the particle, radiation is excited at an angle θ to the path, where cos θ = *c/vn*. v is the particle velocity, *c* is the velocity of light in *vacuo* and *n* is the refractive index of the medium. Thus, the direction of radiation lies in the surface of a cone of semiangle θ, and the electric vector of the radiation is perpendicular to the surface of this cone.

Child's Law. An equation connecting the anode current I_a and voltage V_a for a diode. It is of the form

$$I_a = KV_a^n$$

where K and η are constants. This equation holds under space-charge-limited conditions and can be applied to vacuum and solid-state diodes.

Chi-meson (χ^0). A boson resonance of mass 958 Me V/c². Denoted by 'η'' ('eta-dash'). It has zero isospin and charge, negative parity and positive G-parity. The spin is probably zero.

Chip. A very small semi-conductor having a component (transistor, resistor, etc.) or an integrated circuit.

Choke. An inductor which peresents a relatively high impedance to alternating current. Used in audio-frequency and radio frequency circuits to impede the audio-frequency or radio-frequency signals, or to smooth the output of a rectifying circuit.

Christiansen Effect. The transmission of coloured light when white light is incident on a finely powdered homogeneous and transparent substance, such as glass or quartz, which is immersed in a liquid of the same refractive index. This effect arises because, whereas with monochromatic light complete transparency can be obtained, with white light, because of differences of dispersion in the solid and liquid, refractions will only match for a narrow range of the spectrum.

Chroma. The attribute of a visual sensation which helps us to judge the amount of pure colour, irrespective of the quantity of white or grey present.

Choroid. Refers to the second coating inside the sclerotic of the eye. It is the black, opaque membrane in the interior of the eye.

Chromosphere. The layer of a star's atmosphere that lies betwen the *photosphere* and the *corona*. It is very much less dense than the photosphere. The sun's chromosphere is a few thousand kilometers thick; the temerature increases rapidly from about 4000 ^{0}C near the photosphere to about 500 000 ^{0}C at the base of the corona, with the atmosphere becoming increasingly rarified with height. The solar chromosphere cannot be seen without special equipment, except at a total solar *eclipse*.

Chronon. The time that a photon takes to traverse the diameter of an electron, approximately equal to 10^{-24} seconds.

Chronoscope. An electronic device used for measuring very short time intervals.

Chronotron. A device used for measuring the time interval between events.

Circuit. Refers to a number of an electrical conductors connected together to form a conducting path. When they form a continuous closed path through which a current can circulate. It is said to be closed circuit when the circuit is not closed it is said to be open.

Circuit Breaker. A mechanism for breaking an electrical power circuit under load. Similar in function to a fuse, it has the advantage of being able to be reset immediately. It is used in power stations and high-voltage distribution lines and now often replaces fuses in low-voltage circuits.

Circuit Element. Refers to any device used in an electric circuit. It may be resistance, capacitor, inductance, etc.

Circular Motion. Refers to motion of an object moving in a circle of radius *r* with a speed (constant) *v*.

Circular Polarization (Circularly Polarized Light). Electromagnetic radiation wherein the plane of polarization rotates uniformly round an axis as the light progresses. It is equivalent to the combination of two plane polarized ray out of phase by 90^0 that are moving together.

Cladding. The process of bonding one metal to another in order to prevent corrosion of one of the metals, used in nuclear reactors to prevent corrosion of a fuel element by the coolant and the escape of fission products.

Clark Cell. A voltaic cell that was formerly adopted as a standard of e.m.f. It consisted of a mercury electrode surrounded by a paste of mercury sulphate, the negative electrode is a rod of pure zinc in a saturated solution of zinc sulphate. Its e.m.f. was defined as 1.4345 volts at 15°C.

Clausius-Mosotti Equation. An expression for molar polarization in terms of the dielectric constant ε, the molecular weight M, and density ρ. In SI units

$$P = \frac{\varepsilon - 1}{\varepsilon + 2} \cdot \frac{M}{\rho}.$$

If, in the optical range, ε is replaced by the square of the complex refractive, index, n^2, the *Clausius-Mosotti-Lorentz-Lorenz equation* is obtained. (*See also Lorentz-Lorenz Law*).

Clausius' Virial Theorem. For a system of *n* particles having position vectors r_i ($i = 1, 2, \ldots, n$) and acted upon by forces F_i, the average kinetic energy T of all the particles over time is

$$\bar{T} = -\frac{1}{2} \sum_i F_i \cdot r_i$$

Cleavage. The process of breaking up of a crystal along atomic planes so that smooth surfaces result.

Clock. A mechanical device for measuring the passage of time. Clockwork has two essential components: an energy store (a raised weight or a coiled spring) and an escapement that regulates the release of energy from the store. The earliest recorded escapement was in a giant Chinese astronomical clock (c. 1090 AD). Early European clock (called turret clocks from their usual position in church towers) were crude ironwork with verge escapements, driven by falling weights and recoding time by striking on the hour, but the 14th-century Italian family of Dondi introduced dials on their sophisticated astronomical mechanisms. Late medieval clocks, such as the one (1392) in Wells Cathedral, England, were often embellished with ingenious automata. The innovation of main-springs about 1500 enabled portable clocks to be made. Refinements in the 17th century were the anchor escapement (1671) and the application of pendulums to clockwork. Both resulted in greatly increased accuracy and during the next 150 years clocks successfully challenged *sundials* as the principal domestic time-keepers. Long-case (grandfather) and bracket clocks assumed their modern forms. Carriage clocks were introduced in 1810 by the Frency firm of Breguet. An important advance in the 18th century was the development of an accurate marine chronometer, used by navigators to determine position. Scientific advances in the 20th century have led to the development of clocks powered by the natural vibrations of atoms. The *caesium* clock, first made in 1955, is so accurate that one specific line in the spectrum of caesium is now the standard by which time is defined, replacing the period of the earth's rotation.

Cloud Chamber. A device, invented by C.T.R. *Wilson,* used to study the properties of ionizing particles. It contains a chamber filled with a saturated vapour. The vapour is expanded adiabetically (usually using a piston) to cool it and so make it supersaturated. When an ionizing particle passes through the chamber drops of liquid condense along its trail, thus making the trail visible.

Clock Frequency. Refers to the master frequency delivered by an electronic device, known as a *clock,* at fixed intervals to synchronize operations in a computer, and to monitor and measure properties of the circuits involved.

Clock Pulses. Refers to the regular pulses delivered to the elements in a logic circuit to effect logical operations.

Clocks. The ancient clocks were based on processes taking place at a constant rate sophisticated clocks use periodic processes of constant frequency.

(1) *Pendulum clock.* In pendulum clock the period of a pendulum is a function only of its length and not its mass or initial displacement: accurate to about 0.01 seconds per day.

(2) *Crystal clock.* Is obtained from the crystal clock in which a quartz crystal is allowed to oscillate at about 100.000 hertz by electro-striction: accurate to about 0.001 seconds per day.

(3) *Atomic clock.* In atomic clocks the periodic process is a molecular or atomic event associated with a particular spectral line.

(4) *Caesium clock.* A device in which the frequency is defined by the energy difference between two different states of the caesium nucleus in a magnetic field. The caesium clock is used in the international (SI unit) definition of the second.

Closed Circuit Television. A television system having all the parts (camera, control system, and receiver) physically linked by cables and there are no aerials or open circuits.

Close-Packed Structure. Refers to a crystalline arrangement in which similar atoms, are packed as economically of space as is possible. The two common arrangements are the face centred cubic and hexagonal closepacked structures but combinations of these also occur. For this the essential condition is that each atom shall be symmetrically surrounded by twelve others.

Clutch. A device enabling two rotating shafts to be coupled and uncoupled. Positive clutches have square or spiral jaws to transmit torque without slipping. They can be used where slow speeds and light loads are involved, but normally both shafts have to be at rest for engagement. Friction clutches provide a period of slipping while the shafts are being engaged, and they do not have to be stopped for engagement. The torque is transmitted by friction between attachments to each shaft: in a plate clutch the attachments are flat discs, which press against each other; in a hydraulic clutch they are radial vanes immersed in a fluid. In a motor vehicle a clutch is used between the engine and the gearbox, usually a plate clutch for manual gearboxes and a hydraulic clutch for automatic transmissions.

Coaxial. Having common axis of radial symmetry.

Coaxial Combination of Lenses, Focal Length of. The focal length ξ of a coaxial combination of lenses of focal lengths f_1 and f_2 when they are put a distance d apart is given by

$$\frac{1}{f} = \frac{1}{f_1} + \frac{1}{f_2} - \frac{d}{f_1 f_2}$$

Coefficient of Thermal Conductivity (K). The quantity of heat passing normally per second through a plate of a material having cross-section area of one sq. cm. and one cm. in length, when a steady difference of temperature between the two faces is one degree centigrade.

Coercive Force. The magnetic filed intensity H that D the magnetic flux density in a material to zero. The maximum value of this force is called coercivity.

Coherent. Refers to waves that are in phase both temporally and spatially. The laser is a source of coherent radiation.

Coherent Units. A system of units, such as SI units, in which the quotient or product of any two units gives the units of the resultant physical quantity.

Cohesion. It is the force of attraction between the similar molecules or molecules of the same substance.

Coincidence Circuit. A circuit having input terminals and is designed to produce an output pulse only when both input terminals receive a pulse within a specified time interval.

Cold Cathode. Refers to the cathode of an electronic tube which is caused to emit electrons by having a sufficiently high voltage gradient.

Cold Trap. Refers to a tube, cooled with liquid air or dry-ice (frozen carbondioxide) in acetone, it will condense vapour passing into it.

Collector. The electrode in a transistor through which carriers leave the interelectrode region.

Collimator. A system which produces a beam of parallel light or other radiation, used in spectrometers, telescopes, etc. Also, a small fixed telescope attached to a larger one in order to set the line of sight of the large instrument.

Collision. A collision refers to the mutual action of molecules, atoms, etc., when they encounter one another.

Collision Density. The total number of a specified type of collision that occur per unit time per unit volume of material.

Colorimetry. The science of specifying and repreducing colours as a result of measurement.

Colour. The sensation produced when light of different wavelengths falls on the human eye. Although it is actually continuous, the visible spectrum is usually split into seven major colours: red, orange, yellow, green, blue, indigo, and violet, in order of decreasing wavelength (from about 6.5×10^{-7} m for red light to $4{\cdot}2 \times 10^{-7}$ m for violet). A mixture of all these colours in equal proportions gives white light; other colours are produced by varying the proportions or omitting components. Coloured pigments, dyes, and filters selectively absorb certain wavelengths, transmitting or reflecting the rest. Thus a red book illuminated by white light absorbs all the components of white light except red, which is reflected. This is a subtractive process, since the final colour is that remaining after absorption of the others. Combining coloured light, on the other hand, is an additive process. A mixuture of the whole spectrum gives white light, as will a mixture of lights of three primary colours.

Colour Photography. The recording of *colour* images on photographic film that has three layers of light-sensitive emulsion, one for each of three *primary colours.* Most colour film uses a subtractive reversal process. Coloured light entering the camera first falls on an emulsion sensitive only to blue light. On development of this layer, a black image is formed by deposition of silver where blue light has fallen. The unblackened areas are dyed yellow, the complementary colour of blue, and the silver deposit is removed. Since silver halide emulsions cannot be made insensitive to blue light, there is a yellow filter (to remove blue light) between the blue emulsion and the next layer, which is green-sensitive. This is developed to form a magenta image of the parts where no green light has fallen. The bottom layer is red sensitive and gives a cyan (blue-green) negative image. When white light shines through the three superimposed images, the cyan dye subtracts red where it does not occur in the picture; the magenta subtracts green; and the yellow subtracts blue. The light emerging therefore reconstructs the original picture on a screen, in the case of a transparency, or onto printing paper. Positive colour paper prints directly from negatives that are not to be used for transparencies. These negatives incorporate dyes to correct for varying sensitivity in the printing-paper emulsion. For printing from transparencies, reversal colour paper is used and good colour reproduction is more difficult.

Colour Vision. The ability of the human eye (and that of some other animals) to detect differences in the wavelength of light. The ability is due to the presence in the retina of the eye of cells known a cones, of which three types are believed to exist; one type being sensitive to red light, one to blue, and one to green. Light stimulates one or more of these types of cones in varying amounts depending on its colour.

Coma. A defect of image formed by a lens or mirror, it produces a blurred comet-shaped image.

Compass. A device for determining the direction of magentic north. The magnetic compass, which has been in use as an aid to navigation probably since the 2nd century BC, consists of a magnetic needle balanced on a point, allowing it to pivot freely. The S end of the magnet indicates magnetic N, as shown on a card (called a compass card) marked with the points of the compass and fixed below it. In some magnetic compasses, the entire card pivots, indicating direction against a mark on the fixed housing. Such compasses are often filled with a fluid (usually alcohol) for damping. A more sophisticated kind of compass, used on larger vessels and in aircraft, is the *gyrocompass,* which employs the effect of the earth's rotation on the orientation of a spinning object's axis of rotation. Magnetic compasses are subject to interference from nearby ferous metal objects and fittings, and compasses, must be adjusted to compensate for distortion. Compensation must also be made, in the reading of a compass, for magnetic N not being in the same direction as geographic N in most longitudes. Up-to-date navigation charts mark on a compass rose the annual correction for position that must be allowed in various longitudes.

Complementary Colours. A pair of colours which, when combined give the effect of white light. A large number of such pairs are possible.

Compound Lens. A lens system consisting of two or more lenses. Such a lens system is used as the objective or eye-piece of a telescope of microscope.

Compression. It is region of large pressure along the direction of propagation of a longitudinal wave.

Compressor. A machine for delivering high-pressure gas. They have a wide variety of uses, from small garage machines for inflating tyres to large industrial machines for supplying compressed-air lines. The piston compressor resembles an internal-combustion engine in reverse and is driven by an electric motor or a diesel engine. The centrifugal compressor uses radial vanes to increase the momentum of the gas; they are extensively used in *gas turbines* (driven by the turbine shaft) to supply pressurized air to the combustion chambers.

Compton Effect. Important evidence for the particle or photon nature of radiation is provided by the discovery by A.H. Compton that, if monochromatic X-rays are allowed to fall on a material of low atomic weight, e.g. carbon, the scatered X-rays contain, in addition to those having the incident wavelength, others of somewhat longer wavelength. The change of wavelength is given by

$$\Delta \lambda = \frac{h}{mc} (1 - \cos \theta)$$

where θ is the angle of scattering of a photon and *m* is the rest mass of an electron. If E_p is the energy of the incident photon and E'_p the energy of the scattered photon, then

$$E'_p = E_p \left[1 + \frac{E_p}{mc^2} (1 - \cos \theta) \right]^{-1}$$

Compton-Getting Effect. The sidereal variation of cosmic radiation of extra-galactic origin due to rotation of the earth's galaxy.

Concord. Combination of two or more notes sounded together that produce a pleasing effect on the ear is known as a concord.

Conductance (C). The reciprocal of resistance, units are ohm^{-1} or mho.

Conduction Band. Refers to range of energies in a semi-conductor, the flow of electrons on the application of on external field, in which gives rise to an electric current.

Conduction (1) **(Thermal)** The transfer of heat from a region of high temperature to one of lower temperature, without the transfer of matter. It occurs as a result of the transfer of kinetic *energy* by collisions between atoms and molecules in gases, liquids, and non-metallic solids. In metals, which are the best thermal and electrical conductors, the energy is transferred by collisions between the free electrons that move through the crystal lattice and the rons of the lattice. (2) **(Electrical)** the passage of an electric current through a substance. In metals, the best conductors, it results from the passage of free electrons moving in one direction under the influence of an electric field. In a liquid conductor it is due to the passage of positive ions in one direction and negative ions in the other. In gases it is due to positive ions flowing in one direction and electrons in the other. In *semiconductors* it results from the passage of electrons in one direction and positive holes in the other.

Conductivity, Electrical (σ). (1) The reciprocal of the resistivity or specific resistance of a conductor. (2) The ratio of electric current density to electric field strength.

Counductor, Electrical. A substance that offers little resistance to the flow of electrical current.

Consequent Poles. Refers to Magnetic poles in exess of the usual pair in any magnetized body.

Conservation. The rational use of the earth's resources so that life can be sustained indefinitely. Pressure on both mineral and natural resources, resulting from increased agricultural and industrial activity, has required urgent examination of the consequent effects on the world ecosystem. The prudent use of fossil fuels and minerals and the search for alternatives (*see* alternative energy) are becoming both political and economic necessities. The destruction of many natural habitats has led to the creation of nature reserves and national parks where wildlife can be protected. Legislation can protect some rare species of animals and plants from extinction but international cooperation is needed to protect economically important species (including whales), to conserve fish stocks, and to prevent *pollution* of the atmosphere and the oceans. In most cases the technological means to solve these problems are available but their high costs, or conflict with sectional interests, prevent their implementation.

Conservation Laws. A set of scientific laws stating that the total value of a specified physical quantity remains constant, in a closed system, for any changes that may take place in the system. An example is the *law of conservation of energy,* which says that, in a closed system, energy can neither be created nor destroyed—it can only be transformed from one form into another. Mass (or amount of substance) is another conserved quantity—the *law of conservation of mass*. According to the theory of *relativity,* mass and energy are equivalent (related by Einstein's equation, $E=mc^2$, where E is energy, m mass, and c the speed of light). This leads to the *law of conservation of mass-energy,* in which the total of the mass and energy equivalents remains constant. Other conserved quantities include electric charge, linear momentum, and angular momentum.

Contact Angle. The angle formed between the tangent plane to the surface of a liquid and the tangent plane to the surface of a solid at the point of contact.

Contact Lenses. A removable form of lens worn directly against the eye to replace spectacles for long or short sight or to protect the eye in some disorders of the cornea. Invented in 1887 by Adolf Fick, they were originally made of glass. Modern lenses, developed by Kevin Tuohy in 1948, are made of plastic and can be hard (corneal), gas-permeable (allowing oxygen to permeate the cornea), or soft (hydrophilic).

Contactor. A device for making and breaking an electric circuit designed for frequent use.

Contact Potential. The difference of potential produced when two conductors of different material are placed in contact.

Continuous Wave. Electromagnetic wave produced or transmitted over a period of time and not in short pulses.

Control Rods. Neutron — absorbing rods made of boron, cadmium etc. These can be moved up or down their axis and thus control continuously the rate of nuclear fission reaction in a reactor.

Convection. The transfer of heat within a fluid by means of motion of the fluid. Convection may be natural or forced. In natural convection, the fluid flows by virture of the warmer part being less dense than the cooler part: Thus the warmer fluid rises and the colder fluid sinks under the influence of gravity. In forced convection, some external cause, such as a fan, drives colder fluid into a warmer one, or vice versa.

Convergence (or Convergent Evolution). The development in unrelated animals of similarities resulting from adaption to the same way of life. Thus whales (mammals) and fish, both aquatic, have evolved similar features independently.

Cooling Correction. A correction applied in experiment involving temperature changes as, some heat is lost due temperature difference between the body and the air so a correction has to be applied for that.

Copper Pair. In superconducitivity theory it has been shown that electrons with energy close to the *Fermi energy* can form quasi-molecular pairs called Cooper pairs which can then behave as Bose particles subject to *Bose-Eintein Statistics.*

Copper Loss. Syn. I^2R Loss. Refers to the power loss in watts due to the flow of electric current in the windings of an electrical machine or transformer. It is equal to the product of the square of the current and the resistance of winding.

Corbino Disc. A thin cylindrical disc perforated with a central hole. It is used for making electrical measurements in a magnetic field, with the current flowing radially and the magnetic field perpendicular to the palne of the disc. In this arrangement the *Hall Effect* is 'shorted-out' and there is a maximum magneto-resistance called *Corbino magnetorsistance*.

Corbino Effect. If current is passed from the centre to the circumference of metal disc having its surface at right angles to a magnetic field a current will flow round the circumference.

Core. Refers to either a magnetic material used to concentrate the magnetic lines of force as in a transformer. The central part of a cuclear reactor where the chain reaction takes place.

Coriolis Force. Owing to the earth's rotation a body moving freely upon the earth is deflected to the right in the northern hemisphere and to the lefft in the southern hemisphere. If the velocity of motion is *v*, then (when *v* is not too great) the component of acceleration β at right angles to the direction of motion is

$$\beta = \frac{4\,\pi v\,\sin\theta}{T}$$

where θ is the geographical latitude and T is the period of the earth's rotation. According to *Newton's Laws,* each acceleration may be explained by means of a force and, in this sense, we can speak of the deflecting force of the earth's rotation.' Attention to this force was first drawn by G. G. Coriolis and these forces are sometimes known by his name.

Cornea. The transparent outer layer at the front of the eyeball, through which light enters the eye. Corneal grafts have been used successfully to replace diseased areas of cornea; since the cornea has no blood supply the graft cannot be rejected by blod-borne antibodies.

Corpuscular Theory of Light. A theory that assumes that light travels as particle or corpuscle. Early observations with light like the phenomenon of reflection, refraction, dispersion etc., were explained by it however cannot explain the experimentally observed phenomenon of interference, diffraction etc., and was discarded in favour of the wave theory of light.

Cosmic Rays. A continuous stream of very high-energy particles that bombard the earth from space. The primary radiation consists of *protons* and light nuclei with smaller numbers of neutral particles, such as *photons* and *neutrinos.* These particles collide with atomic nuclei in the earht's atmosphere producing large numbers of elementary particles, known as secondary radiation. One primary particle may produce a large number of secondary particles on colliding with a nucleus. This effect is called a shower. Some cosmic rays are believed to originate from the sun, others from outside the solar system.

Cosmogeny. Refers to the study of the origin and development of the universe or of a particular system in the universe, i.e., the solar systme or a satellite system of a planet.

Cosmology. The branch of astonomy that deals with the evolution, general structure, and nature of the universe as a whole.

Cosmotron. The proton synchotron at Brookhaven, US, that accelerates protons to an energy of 3 GeV.

Cotton-Mouton Effect. When light is passed through certain pure liquids in a direction which is perpendicular to an applied magnetic field, the liquids show double refraction which is much greater than would appear due to teh *Voigt Effect.* It arises when the molecules of liquid possess magnetic and optical anisotropy so that, when a magnetic field in applied, the molecules becomes aligned. It is proportional to the square of the magnetic field strength and is an analogous effect to the *Kerr* (electro-optic) *Effect.*

Coulomb (C). The *SI* unit of electric charge equal to the quantity of electricity transferred by a current of one ampere in one second. Named after Charles de *Coulomb.*

Coulomb, Charles Augustin de (1736-1806). French physicist, who invented the torsion balance and used it to establish *Coulomb's law.* He developed a similar law for magnetic poles. The unit of charge is named after him.

Coulomb Damping. The dissipation of energy of a particle in a vibrating system arising from a resistive force which is of constant magnitude (independent of both displacement and velocity) but which is always in an opposite direction to the direction of velocity of the particle.

Coulomb's Law. The force between two electrically charged bodies is proportional to the product of their charges (q_1 and q_2) and inversely proportional to the square of the distance (d) between them. In free space $F = q_1 q_2/4\pi\varepsilon_0 d^2$, where ε_0 is the *electric constant*. The law is named after Charles de *Coulomb*. The force is known as the *Coulomb* (or electrostatic) *force*. *Coulomb scattering* is the scattering of a charged particle by a nuclues due to the Coulomb force.

Coulmb's Law of Friction. The force required just to move an object along a plane is given by μ N, where *n* is the force normal to the plane and μ is the coefficient of friction, which is independent the area of the surfaces in contact.

Counter. A device used for detecting and counting of particles and electromagnetic radiation.

Couple. Two equal and opposite parallel forces, not acting along a single point, that tend to rotate a body. The movement of a couple (or turning effect) is given by the product of one of the forces and the shortest distance between them.

Critical Damping Resistance (C.D.R.). It is the resistance of a galvanometer at which oscillation just stop. If the resistance is less than C.D.R.; the galvanometer is dead beat ; if it is more than C.D.R.; it is oscillatory.

Crookes Dark Space. As the gas pressure in a discharge tube is gradually diminished, it is found that the glow surrounding the cathode detaches itself, leaving a space around the electrode. This gap is known as the Crookes dark space. At sufficiently low pressures it fills the whole tube. The Crookes dark space is also known as the *Hittorf Dark Space.*

The Crookes dark space often is taken to include the faint red or orange to glow and the *Aston Dark Space* which may occur very close to the cathode, and it is usually only the layer immediately adjacent to the cathode which is totally dark. (*See also Faraday Dark Space*).

Cross Section. Refers to the apparent area that has to be attributted to a particular atom or which is effective in its interaction with an incident bombarding particle. It is a function of the probability of a nuclear reaction that occur with the given incident particle.It is measured in barn; 1 barn = 10^{-24} cm^2.

Cryogenics. Refers to thestudy of materials and phenomenon at very low temperatures close to absolute zero.

Crystat. A vessel in which a low temeperature can be maintained.

Cryotron. A switch that depends on superconductivity. It consists of a wire surrounded by a coil in a liquid helium bath. Both the wire and the coil are superconducting and a low voltage can produce a current in the wire. If a current is also passed through the coil its magnetic field changes the superconducting properties of the wire an switches off the current, thus the presence or absence of a current, in the coil determines the ability of the wire to conduct.

Crystal-Controlled Oscillator. An electromechanical oscillator that uses a mechanically vibrations vibrating quartz crystal. The link between the mechanical vibrations and the electric circuits is provided by the piezoelectric effect of the quartz crystal and can be designed to have a very high degree of frequecy stability.

Crystal Counter. An instrument for detecting and counting subatomic particles that depends on their ability to increase the conductivity of a crystal.

Crystal Detector. A detector depending for its action upon the rectifying properties of some crystals when placed in contact with a metal. It was used extensively in the earliest types of radio receiver, and as a detector and mixer of centimetric waves.

Crystal Difraction. The constructive and destructive interference of waves scattered by the periodic arrangement of electons, nuclei, or field of force in a crystal giving a pattern of discrete spectra.

Crystal Grating. The symmetrical arrangement of the atoms in a crystal in a series of paprallel planes which enables the crystals to act as a three dimensional diffraction grating for X-rays.

Curie (Ci). A unit of radioactivity equal to the amount of an isotope that decays at the rate of $3.7 \times 3.7 \times 10^{10}$ disintegrations per second. Named after Pierre *Curie*.

Curie, Marie, (1867-1934). Polish chemist, renowned for her research into radioactivity. Born Marya Sklodowska, she emigrated to France in 1891, studiied at the Sorbonne, and married (1895)

Pierre Curie (1859-1906), a French physicist. Interested in *Becquerel's* discovery of radioactivity, Marie Curie noticed in 1898 that one particular uranium ore emitted an anomalously large amount of radiation. Realizing that the radiation was cuaused by a new element, she and her husband spend four years isolating one gram of radium salt from eight tons of the ore. The Curies, together with Becquerel, were awarded the 1903 Nobel Prize for Physics and, for her discovery of radium and polonium, she won the 1911 Nobel Prize for Chemistry. Pierre Curie, before joining his wife's work, had discovered the *pezoelectric* effect (1880) and shown that ferromagnetism reverrts to paramagnetism above a certain temperature, now known as the Curie point (1895). He was killed in a road accident; she died as a result of the radiation to which she had been exposed. Their daughter *Irene Joliot Curie* (1896-1956) married the French physicist *Frederic Joliot* (1900-59) in 1926. Working together in the same field as her parents, they were the first to produce radioactivity artificially. For this work they were awarded the Nobel Prize for Chemistry in 1935.

Cureie Point (Temperature). Ferromagnetic materials lose their permanent or spontaneous magnetism when heated above a critical temperature called the Curie point. The value of this critical temeprature depends upon the material. Ferroelectric materials likewise lose their spontaneous polarization above an analogous critical temperature. This upper Curie point must be clearly distinguished from the lower Curie point, below which temperature the poperty of ferroelectricity itself disappears.

Curie's Law. The susceptibility of a paramagnetic substance is inversely proportional to its thermodynamic temperature. The *Curie point* (or Curie temperature) is the temperature above which a ferromagnetic substance becomes paramagnetic. Named after Pierre Curie.

Curie-Weiss Law. If a ferromagnetic susbstance the magnetic susceptibility above the *Curie Point* (*Temperature* is given by

$$X_m = \frac{C}{T-T_c}$$

where C is the Cuire constant = N $\mu^2 / 3\ k_B$ and T_c is the Cuire temperature. N is the number of atoms per unit volume having magnetic moment μ. The law can also be applied to electric susceptibility in ferroelectics.

Currents. Flows of water masses moving in a particular direction. The major ocean currents form part of the general circulatory system in the oceans and are permanent, altheough they may vary with the seasons. They result mainly from the action of the prevailing winds on the sea surface; an example is the North Atlantic Drift. Currents are also induced by the tides, by differences in water density (densities vary with temperature, salinity, and turbidity levels), and by rivers discharging into the sea.

Current Balance. An instrument used for accurately measuring electric current. It involves the measurement of force betwen two conductors carrying the current.

Current Density. (1) Refers to the current flowing through a conductor per unit area. It is expressed in amperes per square meter. (2) (In electrolysis). The current flowing per uit area of electrode.

Curricle. A light two-wheeled carriage usually drawn by two horses harnessed abreast to a pole. Curricles were favoured in the 18th and 19th centuries as rapid and stylish conveyances

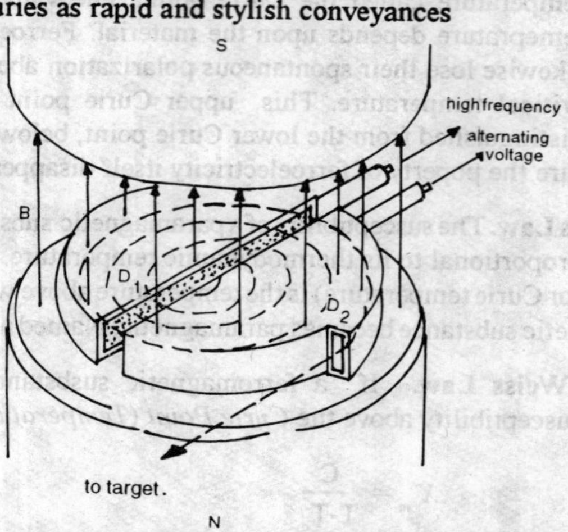

Cyclotron. The charged particles are accelerated in the two D-shaped conductors (D_1 and D_2), which are supported in the magnetic field B.

Cylotron. A type of particle *accelerator* in which charged particles are accelerated in an outward spiral path inside two hollow D-shaped conductors (called dees) placed back to back. A magnetic field at right angles to the plane of the dees causes the particles to move in a spiral and, at the same time, they are accelerated by an alternating electric field applied across the gap between the two dees. When the particles reach the edge of the device they are deflected onto the target. The maximum energy of the particles is about 25 MeV.

D

Dalton's Temperature Scale. In the Kelvin scale the unit of temperature measurement is always the same and is given by

$$t = T - 273.16$$

where t is the temperature on the Celsius scale and T is the temperature on the Kelvin scale.

On the Dalton scale the degree is difined by the relation

$$t = 273 \left\{ \left(\frac{373}{273} \right)^{\tau/100} - 1 \right\}$$

where Γ is the temperature on the Dalton scale. Thus absolute zero on the Kelvin scale is at -273.16°C and on the Dalton scale it is at—∝°C.

Daniell Cell. An electric cell whose positive pole consists of copper immersed in a solution of copper sulphate and whose negative pole consists of zinc in a solution of sulphuric acid or zinc sulphate. It has an almost constant emf of 1.08 volts. Named after J.F.Daniell.

Day. The time taken for the earth to make one rotation on its axis relative to the sun (*solar day*) or relative to the stars (*sidereal day*). The solar day has 24 hours and is the basis of the civil day. The sidereal day, because it takes into account the earth's orbital motion, is 4.09 minutes shorter than the solar day.

Debye Equation for Polarization. The total polarization P for N molecules is given in SI units by

$$P = N \left(\alpha + \frac{p^2}{3k_B T} \right) E$$

where p is the permanent electric dipole moment of the molecule, α is its polarizability and E is the applied field. The first term, due to induced dipoles in the applied electric field, will be present for all molecules; the second term arises in polar molecules having permanent dipoles.

Debye-Scherrer Method (Powder Method). An X-ray diffraction method in which the sample is used in the form of powder. The sample, contained in a thin walled silica tube or stuck to a thin fibre, is rotated in a beam of monochromatic x-radiation. The diffraction pattern is usually recorded on a cylindrical film whose axis is parallel to the axis of the specimen. The pattern consists of a series of arcs arising from the intersection of coaxial cones of radiation with the film. The geometry obeys *Bragg's Law* and the spacing of the arcs cab be used to obtain the lattice parameters of the sample.

Debye-Scherrer Ring. The circular diffraction ring, concentric with the undeflected beam, obtained when a narrow pencil of monochromatic X-radiation is allowed to pass through a mass of finely powdered crystal.

Debye-Sears Effect. A piezoelectric crystal vibrating in a liquid sets up a train of acoustic waves whose nodes occur every half wavelength. It is thus possible to obtain the effect of a diffraction grating by passing a beam of light through the liquid in a direction at right angles to the wave direction.

Debye T^3 Law. At low temperatures the heat capacity of a crystalline solid is approximately proportional to T^3. This can be shown from the Debye Equation.

Debye-Waller Factor. When X-rays are diffracted from a crystal lattice, the intensity is reduced—due to thermal motion of the atoms—by the Debye-Waller factor

$$\exp\left(-1/3 <u^2> < \Delta k^2\right)$$

where $<u^2>$ is the mean square displacement of the atoms from their equilibrium positions and Δk is the change in wave-vector of the X-rays on reflection.

Decibel (dB). A unit used to compare two power levels on a logarithmic scale. It is one-tenth of a bel, but this unit is rarely used. Two power levels P and P_0 differ by *n* decibels when n=/10 $\log_{10} P/P_0$. The unit is often used to express a sound intensity in which case P_0 is usually taken as the intensity of the lowest audible note of the same frequency as P.

Declination, Angle of. Refers to the angle between the magnetic meridian and the geographic meridian at a point on the surface of the Earth.

Declinometer. A device for determining the magnetic declination. It consists of an arrangement by which the angle between the magnetic axis of a compass needle and the direction of some heavenly body can be read directly on a horizontal circular scale.

Decoupling. The removal from a circuit or circuit element of any unwanted a.c. components particularly those with a common power supply. Decoupling is done by using a series inductance or shunt capacitor.

Degaussing. (1) Neutralization of the magnetization of a ship by surrounding it with a system of curent-carrying cables which set up an exactly equal and opposite field. (2) In colour television, the use of a system of coils to neutralize the earth's magnetic field so as to prevent the formation of colour fringes on the image.

Degeneracy. The state of matter when it is at such a high temperature that all the electrons have been stripped from the atoms. The materials is made up of atomic nuclei and electrons and found in neutron stars.

De Haas-van Alphen Effect. Oscillations of the diamagnetic susceptibility of a metal as a function of a strong magnetic field. The effect arises from the quantization of the energy of the electrons and hence the quantization of the electron orbits in the presence of a magnetic field. The latter must be sufficiently strong for electrons to complete several orbits before colliding with imperfections on the surface of the specimen.

Dekatron. A cold cathode scaling tube having ten cathodes arranged in a circle, and associated transfer electodes surrounding the anode. The tubes may be used for switching or for visual display of counts in the decimal system.

Del. *Syn.* nabla. Symbol Δ. The differential operator $i(\delta/\delta\chi) + j(\delta/\delta\gamma) + k(\delta/\delta Z)$, *i, j,* and *k* are unit vectors along the *x*-, *y*- and *z*- axes respectively. The Laplace operator is it ∇^{2t}.

Delayed Neutrons. Neutrons produced from nuclear fission and are not directly formed in the fission process but are produced from exited fission products by beta decay.

Delay Line. A transmission line or any other device which introduce a known delay in the transmission of a signal.

Delay Time. The time taken for a pulse to traverse device or circuit.

Delbruck Scattering. Scattering of photons by the coulombic field of a nucleus is called Delbruck scattering.

Demagnetisation. The process by which a magnet loses its magnetic properties. It can be achieved by hammering or heating the magnet. A magnet is also demagnetized when placed in a coil through which an alternating current is passed.

Dember Effect. The establishment of a voltage in a conductor or semiconductor by illumination of one surface. Hole-electron pairs are produced by the illumination and because, in general, the electrons are more mobile than the holes, a negative charge is established on the non-illuminated side. Hence an electric field is set up, preventing further flow. It is also called the photo-diffusing effect.

Demodulation. The process of separation of the audio (video) signal from the modulated signal wherein the audio signal is superposed on the carrier wave.

Denier. A unit of weight, used to measure the fineness of woven materials equal to 1 gram per 9000 metres.

Density. The mass of unit volume of a substance. In *SI* units it is measured in kilograms per cubic metre; in these units water has a denisty of 1000 kg m^{-3}. *Relative density* (formerly called "specific gravity") is the density of a substance divided by the density of water at 4^0 C. This value is numerically one thousandth of the density. In the c.g.s. system, density (in grams per cubic centimetre) is numerically equal to relative density. The density of a gas (vapour density) is often expressed as the mass of unit volume of the gas divided by the mass of the same volume of hydrogen at standard temperature and pressure.

Depolarizer. A substance, like manganes dioxide that is used to prevent polarization in a voltaic cell.

Derived Units. Refers to units of physical Quantities, other than those of length, mass and time, etc., which are derived from these fundamental units. e.g. the unit of speed is derived in terms of the units of length and time.

Dernier. A unit of weight, used to measure the fineness of woven materials, equal to 1 gram per 9000 metres.

Descartes' Law. When a beam of light passes between two isotropic media, the incident and refracted rays and the normal to the surface lie in the same plane. The incident and refracted rays lie on opposite sides of the surface, and the sines of their angles of inclination bear a constant ratio depending on the media used.

Deviation, Angle of. The angle through which a ray of light deviates from its original path. The angle of deviation of a ray incident on a prism changes with the angle of incidence. It is minimum when the ray passes symmetrically through the prism. It depends on the refractive index μ of the material of the prism and the angle A of the prism by the relation.

$$\mu = \frac{\sin(A+D)/2}{\sin A/2}$$

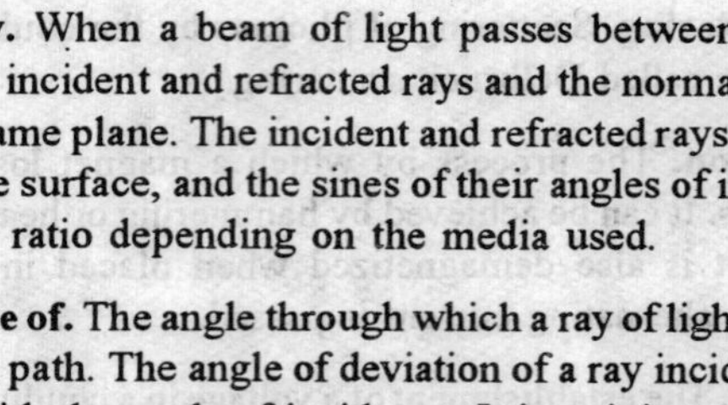

Diaphragm. A device to control the aperture of an optical system so that a desired amour of light enters the system. They also help reduce defects of images resulting from the light incident over a wide part of the opening of the system.

Diathermanous. Substances that allow the radiation to pass through are known as diathermanous.

Dichroism. Some crystals have different absorption for the two polarised components of a doubly refracted beam. The phenomenon is called *dischroism,* and such crystals are known as *dichroic crystals.*

Dichromate Cell. Syn. bicromate cell. A primary cell having poles of carbon and amalgamated zinc immersed in a solution of potassium dichromate ($K_2Cr_2O_7$) in dilute sulphuric acid. A two-fluid form, in which the zinc is in dilute sulphuric acid and the carbon in an aqueous solution of dichromate, the two being separated by a porous partition, has also been used. Its e.m.f. is 2.03 volts.

Dielectric. A substance that acts as an electrical *insulator* and can sustain an electric field. When a voltage is applied across a perfect dielectric there is no energy loss and the electric field strength changes simultaneoulsy with voltage. In the real dielectrics, such as air, ceramics, or wax, which are used in *capacitors*, there is always a small energy loss.

Dielectric Constant (Relative Permitivity). The ratio $\frac{C}{C_0}$ where C is the capacitance of a capacitor with the given substance between its plates as dielectric and C_0 is its capacitance in vacuum.

Diesel Cycle. Diesel was dissatisfied with low efficiency of the Otto engine and was led to investigate certain contrivances whereby the efficiency of the *Carnot Cycle* may be reached. This investigation failed but led to the inventaion of the Diesel engine, which depends upon the idealized Diesel Cycle in the diagram. Pure air is sucked in isobarically (E$\rightarrow$C) and compressed adiabatically (C$\rightarrow$D). A valve is opened and oil is forced in under pressure and burns at the temperature of the system. The supply of oil is regulated so that the piston moves forward and the pressure remains constant (D$\rightarrow$A) the temperature changing from $T_D \rightarrow T_A$. The oil is cut off at A and the gas undergoes adiabatic expansion (A $\rightarrow$ B). The valve is opened at B and the pressure drops to C. C $\rightarrow$ E is a scavenging stroke in which useless gas is forced out and the engine becomes ready for a fresh cycle.

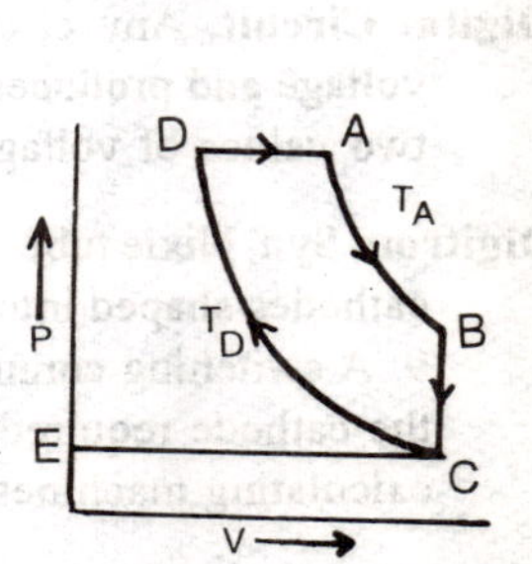

Dieterical Equation. The Dietrici equation is a relation between the pressure of a real gas and its volume, of the form

$$p = \frac{RT}{V-b} \exp\left\{-\frac{a}{TRV}\right\}$$

where a and b are both constants. At low pressures this equation reduces to the *Van der Walls Equation.*

Differential Pulley. A system of two blocks the upper one, consisting of two sheaves of different diameters and cast together to form one piece, and the lower one containg one sheave only. The links of an endless chain fit into the projections on the rims of the sheaves. The mechanical advantage of such a pulley is slightly less than one half its velocity ratio. Its efficiency is less than 50%.

Diffraction. The spreading or bending of light waves as they pass the edge of an object or pass through an aperture. The diffracted waves subsequently interfere with each other producing regions of alternately high and low intensity. This phenomenon, first discovered in 1665 by Francesco Grimaldi (1618-63), can be observed in the irregular boundary of a shadow of an object cast on a screen by a small light source. A similar effect occurs with sound waves.

Diffraction Grating. A (glass) plate having a very large number of closely spaced parallel lines (about 6000 per cm) scrapped across it. Used to produce a spectrum of, and thus analyse, a given light for the various wavelengths present in it.

Digital Circuit. Any circuit that responds to discrete values of input voltage and produces discrete output voltage levels. Unusally only two values of voltage are recognized, as in binary logic circuits.

Digitron. Syn, Nixie tube. A type of cold cathode scaling tube having the cathodes shaped into the form of characters, usually the digits 0 to 9. A switching connection to one side of the power supply selects the cathode required. These tubes are widely used in elecctronic calculating machines and in counting equipment.

Dimensional Analysis. A method of testing or deriving a physical equation in which each term is expressed in the dimensions of mass, length, time, and (if electric or magnetic quantities are present) either charge or current. Each term must have the same dimensional formula if the equation is true. Dimensional analysis is also useful in predicting the behaviour of a full-scale system from a model.

Dimensions of a Physical Quantity. These are the powers to which the fundamental units must be raised, when the quantity is expressed in terms of these units. For example, are a (L^2) is of dimensions two in length, speed ($\frac{L}{T}$) is of dimensions one in length and -1 in time.

Diode. A thermionic valve consisting of two elctrodes, a cathode and an anode. The cathode is heated by passage of an electric current through it and acts as a source of electrons. The anode acts as a collector of electrons. Mainly used as a rectifier and as a demodulator.

Semi-conductor diode. A *p-n* junction across which the current increases exponentially with voltage when biased in the forward direction. There is a very small current in the reverse direction.

Dioptre (D). Unit of power of a lens, it is the reciprocal of focal length of the lens expressed is metres.

Dip. Angle of. *See* inclination.

Dip Circle. An instrument used for measuring the angle of dip at any location. It consists of a magnetic needle mounted so that it can rotate in a vertical plane. The angle is measure on a circular scale.

Dipole Electric and Magnetic. A pair of equal and opposite electric charges or *magnetic poles*. The dipole moment is defined as the magnitude of one of the charges or poles multiplied by the distance separating them. Some molecules have an electric dipole moment due to the preferential attaction of the bonding electrons for one of the atoms. In the case of hydrogen chloride, for example, the chlorine atom gains a slight negative charge and the hydrogen atom a slight positive charge. Measurement of dipole moments provides evidence of the shapes of molecules.

Dirac Equation. A relativistic wave equation for an electron in an electromagnetic field. The dirac Wave-function has four components and the solution requires that, in order that the total angular momentum of the electron be constant, the electron must possess its own intrinsic angular momentum called 'spin'. The existence of the electron spin was deduced empirically by Uhblenbeck and Goudsmit before Dirac's theory was first published.

Discharge, Electrical. The passage of electrical current through a gas at low pressure generally accompanied by luminous glows. The luminous glows generally arise due to recombination of opposite charge ion moving toward their respective electodes.

Discriminator. An electronic circuit that can discriminate between and select desired electrical signals.

Dispersion. The splitting up of a beam of light, which contains several frequencies, on passing it through a substance whose refractive index changes with frequency. When white light is passed through a prism, dispersion takes place and a spectrum consisting of the constituent frequencies is obtained.

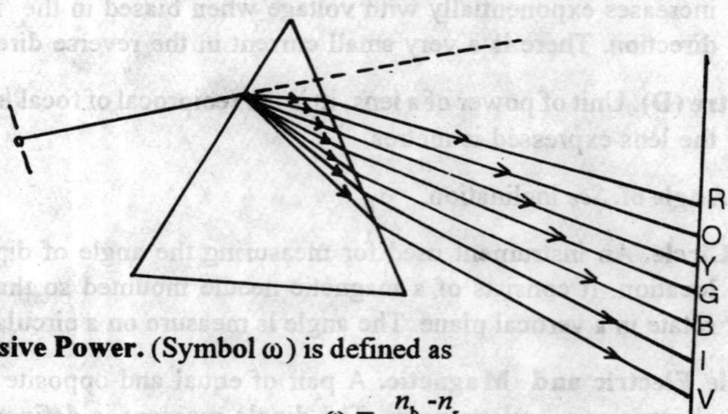

Dispersive Power. (Symbol ω) is defined as

$$\omega = \frac{n_b - n_r}{n_y - 1}$$

where n_b n_r are the refractive indices of the medium for blue, red and yellow light respectively. It is measure of the changes of he refractive index of the medium with wavelength.

Divergent Beam. A beam of light rays that move away from a point. The separation between the rays increases as it moves in space.

Diverging Lens. A lens that renders a parallel beam of light falling onto it into a divergent beam. A concave lens is a diverging lens. Its focal length and power is negative.

Diverging Mirror. A mirror that renders a parallel beam of light falling onto it into a divergent beam. A convex mirror is a diverging mirror.

Domain. Region inside a ferromagnetic material wherein the atomic magnetic moments all point in the same directions. The net magnetization of various domains has been found to be in random directions in an unmagnetized substance. In a magnetized meterial they all point in the same direction.

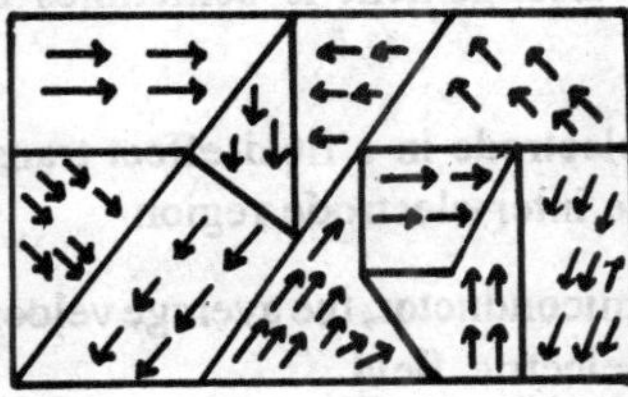

Donnay-Harker Principle. Generalization of *Bravais' Law* for crystal habit by considering the space group of a crystal rather than just the *Bravais Lattice* type, i.e., an extended lattice generalized from point group symmetry. When no glide planes or screw axes are present, the principle is equivalent to Bravais' law.

Doppler Broadening. The broadening of spectral loines on account of the motion of particles emitting or absorbing the radiations corresponding to these lines.

Doppler Effect. The apparent change in the frequency of a wave caused by relative motion between the source and the observer. When the source and the observer are approaching each other, the apparent frequency of the wave increases; when one is travelling away from the other, the apparent frequency decreases. An example of the Doppler effect is the change in pitch of a train whistle as the train passes through a station. The effect can also be observed as a shifting of the wavelength of light from a receding star towards the red end of the spectrum. Named after Christian *Doppler.*

Dose. Refers to energy imparted by an ionizing radiation to living organism and absorbed by it. Maximum permissible dose is the safe upper limit of the dose recommended by International Commission on Radiological Protection. It is measured in rads (10^{-5} J/gm.).

Double Refraction. When a beam of light is incident on certain materials splits it up into two plane polarised beams having their planes of polarisation perpendicular to each other. The two beams have different velocities in the medium. This phenomenon is called *double refraction* .

Doublet. (1) A pair of associated lines in a spectrum having wavelength very close to each other. (2) A compound lens made up of two lenses in contact. The arrangement is sometimes used for minimizing abberrations.

Drain. Refers to the electrode in a field-effect transistor through which carriers leave the inter-electrode region.

Drift Mobility. In a semiconductor, the average velocity of excess minority carriers per unit electric field.

Drift Transistor. *Syn.* A transistor in which the impurity concentration in the base changes across the base region. A high doping level at the emitter-base junction reduces across the base to a low doping level at the collector-base junction. They have a very high frequency reponse.

Drive. A circuit whose output provide the input of one or more other circuits

Driving Point Impedance. The ratio of the r.m.s. value of the sinusoidal voltage applied to two terminals of an electrical network to the r.m.s. value of the current flowing between the terminals as a result of the applied voltage.

Drude's Equation. This relates the specific rotation $[\alpha]$ of polarized light passing through an optically active material to the wavelength λ of the incident light, and may be written as

$$[\alpha]= \frac{k}{\lambda^2 - \lambda^2_0} .$$

k is the rotation constant, λ_0^2 is the dispersion constant and λ is the wavelength of the absorbed light. Within and close to an absorption band the equation does not hold, but in regions of complete transparency an equation involving one or more Drude terms represents, in a very satisfactory way, the variation of angle of rotation with wavelength of the polarized light.

Drude's Theory of Conduction. Drude gave a formula for the specific resistance ρ of a substance, based on classical mechanical considerations, as

$$\frac{1}{p} = \frac{Ne^2}{m}\tau$$

where N is the number of free electrons per unit volume, e, the charge and m the mass of an electron, and τ is the time for each free electron to move between two successive collisions with molecules. The formula leads to values of the mean free path of an electron which are far too large, and the difficulty becomes much worse for conduction at very low temepratures. Application of quantum mechanics has given a new approach to the subject.

Duane-Hunt Law. The high frequency limit for X-rays is proportional to the voltage applied to acclerate the electron beam used to bombard the metal target.

Drum. ***Syn.*** **Magnetic Drum**. A direct access sortage device used in computing which consists of a rapidly rotating cylinder coated with a magnetic iron oxide.

Drum Winding. A type of winding used in electrical machines, consisting of coils, generally former-wound, which are housed in slots either on the outer periphery of a cylindrical core or on the inner periphery of a core having a cylindrical bore.

Dry and Wet Bulb Hygrometer. An arrangement having two thermometers mounted side by side. The bulb of one of these is wrapped in a muslin piece and is kept moist. Relative humidity can be found up from the readings of the dry and wet bulb and the difference between the two.

Dry Cell. A Leclanche cell in which the electrolyte is in the form of paste instead of being liquid. It is used as source of energy in portable electronic appliances and flash lights.

Dulong and Petit's Law. States that the product of the atomic weight and specific heat of a solid element, calledc its atomic heat is constant for all elements and is nearly 6.4.

Dynamics. Branch of mechanics dealing with the study of the behaviour of bodies under the action of forces that changes their motion .

Dynamo. A device that converts mechanical energy into electrical energy. Generally consists of a strong electromagnet between the poles of which is rotated an armature, made up of a number of coils suitably wound. Based on the principle of electro-magnetic induction.

Dyne. The unit of force in the *c.g.s.* system equal to the force that will impart to a mass of one gram an acceleration of one centimetre per second per second.

E

e. The charge of an electron; the natural unit of electric charge. It is equal to 1.60210×10^{-19} coulomb or 4.803×10^{-10} e.s.u. or 1.602×10^{20} e.m.u.

Echelon. A type of grating used for very high resolution spectrosopy. It consists of a number of glass plates having equal thinkness and arranged on a slant in a stepwise formation. The arrangement gives a large value of the gatting element and produces very high spectral orders.

Echo. In Greek legend, a nymph deprived of speech by Hera and able to repeat only the final words of others. Her hopeless love for *Narcissus* caused her to fade away until only her voice remained.

Echolocation. A method by which certain animals can sense and locate surrounding objects by emitting sounds and detecting the echo. Insectivorous bats emit high-frequency sound pulses (12-150 kHz) and, detecting the echo by means of large ears or folds of the nostril, are able to locate their prey when hunting on the wing. Toothed whales and porpoises emit brief intense clicks, enabling them to discriminate objects as small as fine wires. Some shrews and certain cave-dwelling swiftlets also use echolocation as a means of orientation.

Eclipse. The passage of all or part of an astronomical body into the shadow of another. A *lunar eclipse* occurs when the moon can enter the earth's shadow at full moon. The gradual obscuration of the moon's surface is seen wherever the moon is above the horizon. A *solar eclipse*, which is strictly an *occultation,* occurs at new moon, but only when the moon passes directly in front of the sun. The moon's shadow falls on and moves rapidly across the earth. Observers in the outer shadow region (the penumbra) will see a *partial eclipse,* with only part of the sun hidden. Observers in the dark inner (umbral) region of the shadow will see a *total eclipse,* in which the sun's disc is completely but briefly obscured; the *corona* can, however, be seen. If the moon is too far away totally to cover the sun, an *annular eclipse* is observed, in which a rim of light is visible around the eclised sun.

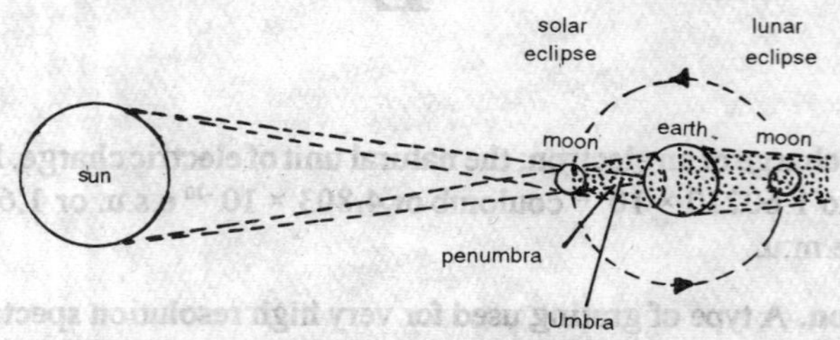

Ecliptic. The great circle in which the plane of the earth's orbit around the sun meets the celestial sphere. As a result of the earth's orbital motion, the ecliptic marks the apparent path of the sun across the celestial sphere, relative to the background stars, over the course of a year.

Eddy Current. An electric current induced in a conductor that experiences a changing magnetic field. In *electric* generators, motors, and transformers, energy is lost by unwanted eddy currents, either producing heat or opposing motion. In these devices eddy currents are reduced by laminating the iron cores to increase their resistance. Eddy currents are useful in some applications, such as eddy-current heating, mechanical damping and eddy-current brakes, and electricity meters.

Edser and Butler's Bands. Dark bands, with constant frequency separation, in the spectrum of white light which has traversed a thin parallel sided plate of a transparent material.

Efficiency (η). Refers to the ratio of the ouput to the input for an engine or system. For a machine, it may be defined as the force ratio divided by the distance ratio.

Efficiency of an Accumulator. It is the ratio of the energy, measured in watt-hour, received from the cell during discharge to the energy supplied to it in the process of charging.

$$\text{Thus, Efficiency} = \frac{\text{output of energy}}{\text{Input of energy}}$$

Echrenfest's Adiabatic Law. For a virtual infinitely slow, alteration of the coupling conditions in a system, the quantum numbers do not change. The law is applicable both in *Bohr's Theory* and in quantum mechanics. The converse states that only such magnitudes as remain invariant for adiabatic changes can be quantized.

Ehrenfest's Theorem. If the position and momentum of a classical particle are replaced by their quantum-mechanical expectation values then the motion of a quantum mechanical wave packet satisfying *Schrondinger's Equation* will be equivalent to that difined by the classical equations of motion for the particle, provided any potentials acting upon the particle do not change over the dimensions of the wave-packet.

Ehrenhaft Effect. A helical-type movement of fine particles in the lines of force of a magnetic field during irradiation by light. The movment is similar to that of electrically charged particles in an electric field and, although the motion was originally thought to be due to 'magnetic charges' on the particles, it is caused by radiation pressure.

Einstein's Equation. Refers to the equation, maximum (K.E.) = $hf—\omega$ for the kinetic energy of electrons which are emitted in photoelectric effect, f the frequency of incident radiation and ω the work function of the photomaterial upon which the radiation is incident.

Einstein Relation (for Mobility). A relation connecting the diffusion coefficient D and mobility μ of an ion (or charged particle) of the form

$$\mu k_B T = eD$$

where e is the charge on the particle.

Einstein's Equation for Specific Heat. It can be deduced from the classical atomic model that the specific heat of a solid at constant volume is connected to the gas constant R by the equation.

$$C_v = 3R$$

This is not true, however, as all specific and hence atomic heats are temperature-dependent. By application of quantum-mechanical methods to this problems, Einstein arrived at the equation

$$Cv = 3R\left(\frac{hv}{k_BT}\right)^2 \frac{\exp hv/k_BT}{(\exp hv/k_BT-1)^2}$$

where v is the *Einstein Frequency,* the resonance frequency of the oscillator.

For most elements hv/k_BT is sufficiently small for C_v to equal 3R. Deviations from *Dulong and Petit's Law* are due to a high value of v or to low temperatures.

Einstein's Law. In the emission of electrons from metals by the incidence of radiation, if W is the work required to pass the electron through the surface, the maximum energy of the emitted electron is given by

$$1/2mv^2 = hv - W$$

where m is the mass and v the velocity of the electron, v is the frequency of the radiation used and h is *Planck's Constant.* W is known as the 'work function' of the metal.

Otherwise expressed

$$V_e = hv - hv_o$$

where V_e is the magnitude of the retarding potential required to decrease the velocity of the electron to zero, and $v_o = w/h$ is the minimum frequency which causes photoelectric emission from the given surface and is known as the 'photoelectric threshold'.

Einstein's Principle of Relativity.A theory of dynamics involving a space-time continuum and based on two major postulates:

(1) Unifom motion of translation cannot be measured or detected by an observer stationed on a system of coordinates for measurements confined to that system.

(2) The velocity of ligh in space is constant and independent of the relative velocities of the source and the observer.

These postulates apply to the Special Theory of Relativity where inertial frames move with constant velocities with respect to each other.

Elastic Collision. A collision in which the momentum as well as the kinetic energies of the colliding bodies are conserved on collision

Elasticity. In physics, the ability of a body to return to its original shape after being deformed. The deforming force is known as a *stress,* the resulting deformation is the *strain*. A body is elastic only below a certain stress. Above this point, known as the *elastic limit,* the body is permanently deformed. A substance permanently deformed by any stress is said to be plastic.

Elastic Limit. When a body is subjected to an external stress, a strain is produced in the body. For short stresses, the train produced is proportional to the applied stress. If the stress is increased, beyond a certain limit Hooke's law is no longer obeyed. With further increase in stress, a stage is reached when the system no longer returns to its original shape (and size) when the deforming stress is removed. The limit above which this happens is known as the elastic limit of the body.

Elastic Modulus. The ratio of the *stress* on a body obeying Hooke's law to the *strain* produced. The strain may be a change in length (Young's Modulus) or a change in volume (bulk modulus).

Electret. A permanently electrified dielectric that has a positive charge at one end at a negative charge at the other. It has permanent dipole moment.

Electrical Engineering. The branch of engineering concerned with the grneration, transmission, distribution, and use of electricity. Its two main branches are power engineering and *electronics* (including telecommunications). Electrical engineering emerged in the late 19th century with the mathematical formulation of the basic laws of electricity by James Clerk *Maxwell,* followed by the development of such practical applications as the Bell telephone, Edison's incandescent lamp, and the first central generating plants. Electrical power engineers design generators, *power stations,* and *electricity supply* systems as well as *electric motors* and transport and traction systems. Electrical engineering is an applied science involving mathematical skills and a knowledge of physics, in addition tothe basic engineering subjects. In the UK the professional body is the Institution of Electrical Engineers (founded 1871, incorporated 1921).

Electric Braking. A method of braking an electric motor by causing it to act as a generator. It has a particular application in electric traction.

Electric Bell. A device in which to and for movement of a small hammer on a metallic vessel produces the sound of a bell. It consists of an electromagnet which attracts the hammer when a button is pressed. The attraction causes the circuit to break bringing the hammer back to its original position. The circuit closes again, and the process is repeated many times giving an almost continuous sound.

Electric Car. A car driven by one or more electric motors, which are powered by batteries. Because of diminshing oil reserves and the pollution problems associated with the *internal-combustion* engine, the search for an effective electric car has been intensified in recent years. Electric cars are not new; Queen Alexandra owned an American *Columbia* electric car in 1901 and electric milkfloats and delivery vans have been widely used for short journeys in towns since the 1930s. The basic problem is that even a light car using 280 kg of expensive nickel-cadmium batteries with regenerative braking, only has a 50 - 60 miles range. Electric motors are several times more efficient than petrol engines but it was not until 1990 that General Motors announced plans to produce an electric car—the *Impact*—to compete with petrol cars. Two lines of research are being pursued: improving accumulators and making workable *fuel cells*. However, unless the batteries are charged by electricity generated by *nuclear power* or *alternative energy* sources there will be no net reduction in global pollution. Several commuter cars using special lead accumulators have been produced, with top speeds around 30 mph (44 km per hour), including the unsuccessful Sinclair C5 introduced in 1985 at under £400. The hybrid electric car carries its own petrol-engine-driven generator.

Electric Cell (or Voltaic Cell). A device that produces an *electromotive* force and delivers a current as result of a chemical reaction, *Primary cells,* such as the *Daniell cell* and the *Leclanche cell,* are not reversible. *Secondary cells,* however, can be recharged.

Electric Charge. A property of certain elementary particles that causes them to undergo *electromagnetic* interactions. The magnitude of the charge is always the same and is equal to 1.6021 × 10^{-19} coulombs (although quarks, if they exist, would have a fractional charge). Charge is of two kinds, arbitrarily called positive and negative. Like charges repel each other and unlike charges attract each other. The force between them can be regarded as being generated by an exchange of virtual photons between the two particles. On a large scale, charge is always due to an excess or deficiency of electons compared to the number of protons in the nuclei of a substance.

Electric Constant (ε_0). A constant that appears in *Coulomb's law* when expressed in SI units. Its value is 8.854 × 10^{-12} F m^{-1}. It is also known as the absolute permitivity of free space.

Electric Field. The pattern of the lines of force that surround an electric charge. The field strength at any point is inversely proportional to the square of the distance of that point from the charge. Any other charge placed in this field experiences a force proportional to the field strength and to the magnitude of the introduced charge. The force is attractive if the charges are opposite and repulsive if they are alike.

Electric Field Strength (Also called Electric Intensity) ($\vec{E}$). The force experienced by a unit charge placed at a point in an electric field is known as the electric intensity at that point. The force experienced by a charge q placed in an electric field, of intensity $\vec{E}$ is $q\vec{E}$.

Electric Flux (ϕ). Refers to the number of lines of force of an electric field through a given perpendicular area. The electric flux through an area *ds* can be expressed as $\vec{E}\,\vec{ds}$, where $\vec{E}$ is the electric field.

Electric Generator. A device for converting mechanical energy into electricity, usually by *electromagnetic* induction. A simple electromagnetic generator, or dynamo, consists of a conducting coil rotated in a magnetic field. Current induced in this coil is fed to an external circuit by slip rings in an alternating-current generator (called an alternator) or by a commutator, which rectifies the current, in a direct-current generator. Most of the electricity from *power stations* is produced by alternators as three-phase *alternating current*, i.e. there are three windings on each generator, which can have an output of hundreds of megawatts, producing three separate output voltages. For transmission there are three live conductor wires with a common neutral wire. This three-phase system optimizes generator design and minimizes transmission losses, if the three loads are balanced. Generally, all three phases are supplied to large factories but the supply is split to single phase for homes, shops and offices. Small generators driven by Diesel engines are used for emergency supplies in factories, hospitals, etc. In recent years interest has grown in generators using renewable energy sources .

Electric Image. A charge Q placed at a certain distance from a conducting surface has the same effect as if we have placed a charge and its image having a charge—Q_2 which, for a plane conducting surface, is situated as far behind the surface as the object is in front of it.

Electricity. The phenomena that arise as a result of *electric charge*. Electricity has two forms: *static electricity,* which depends on stationary charges, and *current electricity,* which consists of a flow of charges, specifically *electrons.*

Static electricity, in the form of an attractive force between rubbed amber and pieces of straw, etc., was known to the ancient Greeks but the word electricity was coined in the 16th century by William Gilbert (from the Greek *elektron,* amber). The distinction between positive and negative electricity was made at the beginning of the 18th century, but did not acquire a theoretical basis until the discovery of the electron in 1897 by J.J. Thomson.

Current electricity was first demonstrated by *Volt* in 1800 and investigated by *Ampere* during the next 25 years. *Oersted's* discovery (1820) that a magnetic needle was deflected by an electric current inspired *Faraday* to investigate the relationship between electricity and magnetism, which led to the discovery of electromagnetic induction, the *electric generator,* and the *electric motor.* The theory of electromagnetism was elucidated by Clerk *Maxwell* in the mid 1850s. The use of electricity as a source of energy developed for heating, lighting, and motive power in the 20th century. Our dependence upon it is now taken so much for granted that life seems almost unimaginable without it.

Electric Motor. A device that converts electrical energy into mechanical energy usually for driving machinery. Clean, quiet, and efficient (75-95%), they range from tiny models using less than a watt to those using several megawatts to drive large pumps. They work on the principle that a current-carrying conductor in a magnetic field experiences a force and that when two electromagnets are placed close together the two magnetic fields force them apart. The simplest motor uses this principle to turn a single coil of wire (the armature or rotor) between the poles of a permanent magnet. Practical motors use a stationary winding (stator) in place of a magnet. Most motors work on alternating-current (ac). In the induction motor, current is fed to the stator, which induces a current in the rotor; interaction between the magnetic field of the stator and the induced rotor current causes the rotor to rotate. In the synchronous *ac* motor, current is also fed to the rotor (through slip rings) and the rate of rotation is proportional to the supply frequency. In direct-current (dc) motors, current is also fed to both rotor and stator with the two either in series (series wound) or in parallel (shunt wound). In a *dc* motor current is fed to the rotor through a commutator.

Electric Polarization (P). Refers to relative displacement of positive and negative charges in an atom or dielectric when it is placed in an electric field.

Electric Potential. Symbol V. The electric potential at a point in an electric field is equal to the work required to bring unit positive electric charge from infinity to that point. It is measured in volts. If work of 1 joule is required to move a charge of 1 coulomb to the point its potential is 1 volt.

Electric Susceptibility (X). The ratio $\frac{P}{E}$ where P is the polarization produced in a dielectric on being subjected to an electric field E.

Electrolytic Capacitor. A capacitor having one plate is a thin metal plate coated with a thin oxide layer immersed in a suitable electrolyte (H_2SO_4). They provide a high capacitance with a small size of the capacitor.

Electomagnet. A magnet obtained by windig a coil of wire around a piece of soft iron. It behaves as a magnet so long as acurrent pass through the coil.

Electromagnetic Induction The production of voltage in an electrical conductor when it is in a changing magnetic fied or if it moves in relation to a steady magnetic field. The direction of the induced *electromotive* force opposes the change or motion causing it. Since a current-carrying conductor itself induces a magnetic field, if the current changes *self-inductance* occurs, opposing, the current change. *Mutual inductance* occurs between two adjacent conductors that carry changing curents.

Electromagnetic Interaction. An interaction that occurs between those elementary particles that possess an electric charge. The interaction can be visualized as the exchange of virtual photons between the interacting particles. The electromagnetic interaction is 200 times weaker than the *strong* interaction but 10^{10} times stronger than the *weak* interaction.

Electromagnetic Radiation (Electromagnetic Wave). They consist of oscillating electric and magnetic fields at right angles to each other and to the direction of propagation of the wave. Such waves require no medium for propagation and can travel through vacuum as well. Light is a form of an electromagnetic wave. They travel through vacuum with a speed of nearly 3×10^8 m/s. Their Frequency ,f, and wave length, λ are related as $C = f\lambda$, where c is the speed of light. The energy of electromagnetic radiation is given by E= hf, where h is called Planck's constant.

Electrometer. A device for measuring potential differences between two points in a circuit. It draws very little current in the process.

Electromotive Force (emf). The electrical potential difference or voltage between two points in an electric circuit. It causes the movement of charge that constitutes an electric current, providing a limited analogy to mechanical force and motion.

Electron. A stable negatively charged elementary particle, classified as a *Lepton,* with mass $9.109\ 56 \times 10^{-31}$ kilogram and spin 1/2.

Electrons are responsible for almost all commonly observed electrical and magnetic effects and, since they orbit the nucleus in atoms, are also responsible for most chemical processes. A *free electron* is one that has become detached from an atom. An electric current passing through a metal or low-pressure gas consists of a flow of free electrons; a current of 1 ampere is equivalent to a flow of 6×10^{-18} electrons per second. Electrons can also have wave characteristics.

Electron Diffrction. The *diffraction* of a beam of electrons as it passes through spaces, the widths of which are comparable to the wavelength of the electrons. Electrons can be diffracted, for example, by the spacing between particles in a crystal lattice, an effect governed by *Bragg's law*. Electron diffraction demonstrates the wave aspect of electrons since diffraction is specifically of surfaces, films, etc.

Electronics. The study of devices that control and utilize the movement of *electrons* and other charged particles. Originating with the iinvention of thermionic valves and their use in radios and record players, it expanded rapidly during World War II to include radar, missile guidance systems, and the first electronic *computers*. The

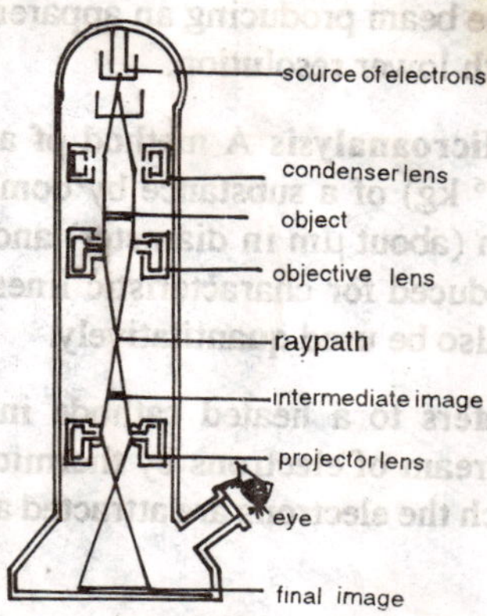

replacement of bulky *thermionic* valves by *semiconductor* components, such as *transistors*, made possible high-speed digital computers and the miniaturized communications and control systems used in spacecraft. The *integrated* circuit was a further step towards more compact and reliable equipment. The latest development is the use of computers to produce tiny wafers of silicon functioning as microprocessors.

The social impact of elecronics has been immense, with the development of television, communications satellites, the computerization of office and factory systems, and the widespread use of pocket calculators.

Electron Microscope. (1) The beam of electrons in the transmission electron microscope is focused in a similar way to light in an ordinary microscope. (2) A type of microscope in which a beam of electrons is focused by means of magnetic and electrostatic lenses onto a specimen and scattered by it to produce an image. The *de Broglie* wavelength of high-energy electrons is very much less than that of light. Therefore, both the *resolving power* and the magnification of an electron microscope are much greater than can be obtained with an optical microscope. Typically an electron microscope can resolve two points 10^{-9} metre apart and produce magnifications of up to a million. In the *transmission electron microscope* only very thin specimens can be used: the image appears two-dimensional but a high resolution can be obtained. In the *scanning electron microscope* the specimen, which can . be of any thickness, is scanned by the beam producing an apparently three-dimensional image but with lower resolution.

Electron Probe Microanalysis A method of analysing a very small quantity (10^{-16} kg) of a substance by bombarding it with a fine electron beam (about μm in diameter) and examining the X-ray spectrum produced for characteristic lines of the elements. The method can also be used quantitatively.

Electron Gun. Refers to a heated cathode in a vacuum tube, that produces a stream of electrons by thermionic emission, with an anode to which the electrons are attracted and focussing electrode (s).

Electron Lens. A combination of electric and magnetic field used for focussing a beam of electron similar to an optical lens focussing a beam of light.

Electron Optics. It is the study of the behavious and control of an electron beam in magnetic and electrostatic fields.

Electron Stains. Substances like posphotungstic acid, osmic acid, silicotungstic acid, and phosphomolybdic acid having high electron-scattering power and can be used with the electron microscope in the same way as staining media in optical microscopy.

Electron Telescope. A type of telescope where in ultraviolet and infrared radiation can be converted into visible image. The intensity of faint visible radiation can be increased. An accurate visible image is obtained.

Electron Temperature. The electron temperature of the plasma is that temeprature at which gas molecules would have the same average kinetic energy as the electons of the plasma.

Electron Tube. A device where in the movement of electons between two electodes, through a gas or a vacuum, takes place in sealed or continuously exhausted envelope.

Electronvolt (eV). A unit of energy, widely used in nuclear physics, equal to the increase in the energy of an electron when it passes through a rise in potential of one volt. $1\ eV = 1.6 \times 10^{-19}$ joule.

Electro-optics. Refers to the study of the changes in the optical properties of a dielectric produced by the application of an electric field.

Electrophoresis. It is the process of the migration of fine particles of solid suspended in liquid to the anode (*anaphoresis*) or cathode (*cataphoresis*) when an electric field is applied to the suspension.

Eelectroplating. The process in which surface of one metal is coated with another, either for protection or for decoration or both by electrolysis.

Electroscope. An electostatic instrument that detects electric charge or radiation. In a gold-leaf electroscope, the deflection of two suspended gold leaves increases with charge, The Quartz-Fibre Electroscope (QFE), which has a quartz fibre instead of gold leaves, is commonly used to detect radiation. The presence of a radioactive source ionizes the air and causes charge to leak away from the initially charged fibre.

Electrosmosis. *Syn.* electroendosmosis. Refers to the passage of an electrolyte through a membrance or porous partition under the influence of an electrometer.

Electrostatic Deflection. A method used for deflecting an electron beam making use of the electrostatic fields produced between two meal electodes.

Electrostatic Generators. Machines that use mechanical or other energy ot separate electric charge, creating an electric potential. They are used to creat strong electrostatic fields, particularly in high energy nuclear physics for accelerating charged particles to bombard atomic nuclei.

Electrostatic Induction. The deposition or separation of positive and negative charges in a substance when a charged body is brought near it, for example, when a positively charged body is brought near a rod, the end of the rod near the charged body acquires a negative charge by induction and a positive charge is induced at its far end.

Electrostatic Precipitation. A mechanism used to control the pollution of air (or other gases). The gas having solid or liquid particles, is subjected to a unidirectional electrostatic field, such that the particles are attracted to, and deposited upon, the positive electrode.

Electrostatics. (1) Refers to the study of the effects of static charges in their neighbourhood. The study of the electric fields and potentials produced by them. (2) *See* static electricity.

Electrostatic Separation. The use of a strong electric field to separate substances with different electrical properties. The method is used, for example, in separating out iron ores from a mixture of minerals, and to clean air in chimneys by precipitating charged smoke particles.

Electrostatic Shield. A conducting substance which protects a given apparatus against electric fields. It consists of a hollow conductor that completely surounds the apparatus to be shielded because the electric field inside a closed conductor is zero, external electric fields do not influence the given apparatus.

Elliptical Polarization. (Ellipticall polarised light): Electromagnetic radiation in the visible region that can be resolved into two planes-polarized vibrations of same frequency but having unequal amplitudes, which are out of phase by 90⁰.

Elster and Geitel Effect. In the presence of a gas, a heated conductor assumes a charge either positive or negative. In a vacuum the charge is always netagtive.

Emissivity. The ratio of the emissive power of a body to the emissive power of a perfectly black body at the same temperature.

End Correction (Electricity). Refers to the correction required in the reading of balance point in a slide wire bridge in orcder to correct the error due to the resistances at the ends of the bridge wire. Generally the end resistances are very small and are neglected.

End Correction (Resonance Tube). In a resonance tube, the reflection of sound waves takes place slighly above the open end not at the open end itself. The length measured from the open end, therefore, requires a correction known as the end correction. It is equal to 0.3 D, where D is the diameter of the resonance tube.

Endothermic Process. A process in which heat is absorbed.

Energy. A property of a system that enables it to do work, i.e., to move the point of application of a force. The several forms of energy can be converted into each other under appropriate conditions. *Kinetic* energy is energy of motion, whereas potential energy is stored energy, for example, the energy stored in a body by virtue of its position in a gravitational or electric field. Other forms of energy include thermal energy (the kinetic energy of the atoms and molecules in a body), chemical energy (the potential energy stored in the chemical bonds between atoms in a substance), nuclear energy (the potential energy stored in the atomic nucleus), and radiant energy the energy associated with electromagnetic waves).

As a consequence of the special theory of relativity, mass (m) has also to be regarded as a special form of energy (E), in accordance with the equation $E = mc^2$, where c is the speed of light. Thus the production of nuclear energy involves a loss of mass in the fuel. However, the sum of the mass and the energy is conserved. Energy is measured in joules (SI units), calories or ergs (c.g.s. units), kilowatt-hours or British thermal units (Imperial units).

Energy sources: man's first use of energy (other than that of his own body or the body of animals) came with the discovery of fire. For combustion is a process in which chemical energy is converted into thermal energy. The first fuel was wood; but fossil fuels (asphalt, coal, oil, natural gas) have been in use for about 8000 years. However, it was the *industrial* revolution and the later advent of motorized transport that brought explosive increases in the demand for energy and consequently for fossil fuels. These demands have steadily increased during the 20th century, especially with the spread of technology throughout the world. The total world energy consumption is of the order of 4×10^{20} joules per year, nearly 90% of which is provided by fossil fuels. Concern both about the exhaustion of fossil fuel reserves and the 2.5×10^{10} tonnes of CO_2 they produce each year is directing attention to *alternative* energy sources as well as nuclear energy.

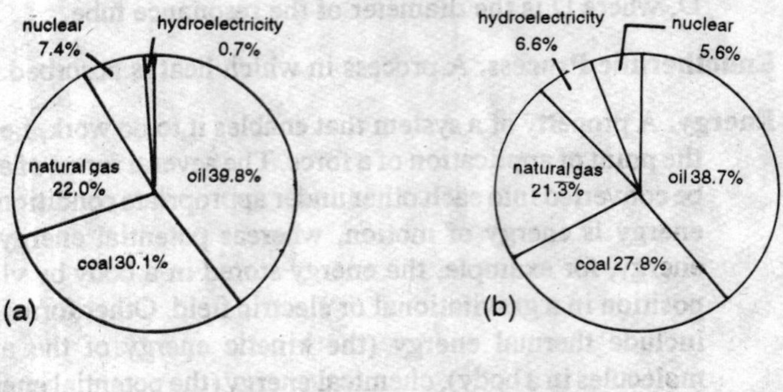

Energy Primary sources (1989): (a) U.K. (b) World.

Energy Band. A concept used to explain the electrical properties of solids, particularly *semiconductors*. Electrons need to have a minimum energy, E_v, to free them from the atoms of the solid. Below this energy they are said to be in the valence band. To be able to move through the solid, so that they constitute a current, they need to be above another energy level, E_c, in the so-called conduction band. In an insulator, E_c is significantly greater than E_v and ther are effectively no electrons in the conduction band. n a conductor, there are many electrons with energy greater than E_c, and E_c may be less than E_v, i.e. the conduction and valence band are very close or overlap. A pure semiconductor at *absolute zero* behaves like an insulator, but the difference between E_c and E_v is small. With an increase in temperature (a measure of the average energy of the particles in the material), some electrons enter the conduction band. Also, the presence of impurities in a semiconductor crystal brings electron of an intermediate energy level with it, since E_v and E_c vary from one material to another. Depending on the type of impurity, the number of electrons in either the conduction band or the valence band at a particular temperature, will increase, changing the electrical properties of the solid.

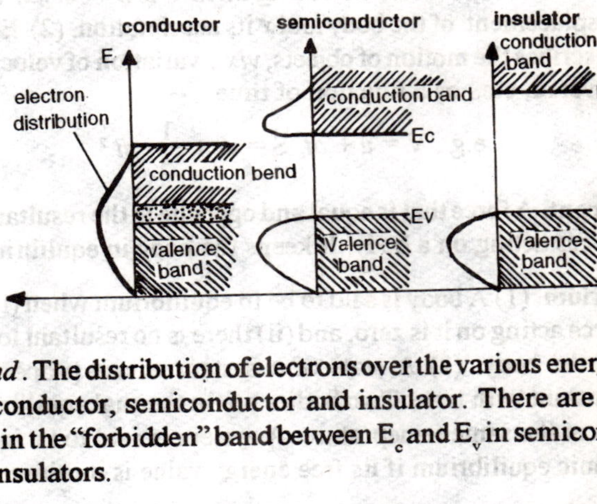

Energy Band. The distribution of electrons over the various energy levels in a conductor, semiconductor and insulator. There are no electrons in the "forbidden" band between E_c and E_v in semiconductors and insulators.

Enthalpy (H). It is given by H=U + *pv*, where U is the internal energy, *p* the pressure, and *v* is the volume. It is also a state function.

Entropy (S). A measure of the degree of disorder of a system. Change of entropy ΔS for a reversible process is $\Delta S = -\dfrac{\Delta Q}{T}$, where ΔQ is the heat supplied to the system at temperature T. The entropy of a system never decreases. It increases for an irreversible process and remains constant for a reversible process.

Epidiascope. An optical projector for throwing an enlarged image of either an opaque object or a transparent object upon a screen. It is used to illustrate lectures.

Epithermal Neutron. A neutron having energy just above thermal energies, generally in the range from 10^{-2} to 10^{2} eV (1.6×10^{-21} to 1.6×10^{-17} joule. Epithermal neutrons have energies of the same order as the energies of chemical bonds.

Equalization. In electronics. the introduction of networks which compensate for a particular type of distortion over the frequency band required and thus reduce distortion in a system.

Equations of Motion. (1) Equation that describes the dependence of acceleration of a body on different variables e.g., the equation of motion of a particles executing S.H.M. is $\ddot{x} = -kx$, were x is the displacement of the body and $\ddot{x}$ its acceleration. (2) Equation that describes the motion of objects, *vix.*, variation of velocity, distance covered, etc., as a function of time

$$\text{e.g., } V = u + at,\ S = ut + \frac{1}{2}at.^2$$

Equilibrant. A force that is equal and opposite to the resultant of various forces acting on a body. It keeps the body in equlibrium.

Equilibrium. (1) A body is said to be in equilibrium when (i) sum of the force acting on it is zero, and (ii) there is no resultant torque acting on the body. (2) A state of a system whose properties remain constant with time. Two bodies are in thermal equilibrium if they are at the same temeprature. A system is said to be in thermodynamic equilibrium if its free energy value is minimum.

Equipartition of Energy, Principle of. The total energy of a molecule is equally distributed among the various degrees of freedom of the molecule. For a system at temperature T, the average energy per degree of freedom is 1/2 kT, where k is called the Boltzman's constant.

Equipotential. An equipotential surface is a surface drawn so that all point on it have the same potential. It cuts the lines of force orthogonally, and no work is done against electrical forces moving a small charge in any direction in the surface.

Equivalent Circuit. An arrangement of simple circuit elements which has the same electrical characteristics as a more complicated circuit or device under specified conditions.

Equivalent Focal Length. The distance from a principal point to its corresponding principal focal point.

Equivalent Length of a Magnet. The poles of a magnet are not situated at its ends. An equivalent length ^{2l} can be chosen such that ^{2l} times the pole strength is equal to the moment. This length is commonly about 5/6 of the geometrical length for simple bar magnets.

Equivalent Network. An electrical network that can be connected in place of another without changing in any way the conditions in the other parts of the system .

Equivalent Positions (Points). Refers to a complete set of points in any given space group that is obtained by performing the symmetry operaions of the space group on a single point (x, y, z).

Equivalent Resistance. The resistance that, when concentrated at a point in electrical circuit, would dissipate power equal to the total of various smaller resistances at different points in the circuit.

Equivalent Sine Wave. A sine wave with the same root-meansquare value as the given wave and also the same fundamental frequency.

Erg. The unit of energy in the *c.g.s.* system equal to the work done when a force of one dyne acts through a distance of one centimetre. 1 erg $= 10^{-7}$ joule.

Ergon. A quantum of energy of an oscillator and is equal to the product of the frequency of oscillation and the Planck constant.

Ergonomics. The study of the psychological and physical factors that can be usd to improve the design of both machines and systems for human use. In the USA the term human-factors engineering is widely used and in other parts of the world ergonomics is known as psychological engineering. The study finds application in many fields, including astronautics, the design of aircraft and cars, and a wide variety of industrial processes. Although it is a young science, having largely grown up since World War II, most university engineering courses now include it.

Error Equation. The equation

$$y = h\pi - \frac{1}{2} \exp[-h^2(x-a)^2].$$

Esaki Effect. A tunnelling effect which occurs in heavily doped *p-n* junctions with narrow space-charge layers. If the voltage across the junction is gradually increased, the currant increases linearly to a peak value and then decreases to a broad minimum before increasing again. The peak in current arises when the voltage bias causes electrons in the conduction band on the *n*-type side to be at the same energy level as unoccupied states (holes) in the valence band on the *p*-type side. This results in tunnelling of electrons to these unoccupied states. Further voltage bias raises the bottom of the conduction band of the *n*-type material above the valence band of the *p*-type and then tunnelling ceases. The effect has application to the tunnel diode.

Escape Velocity. The initial velocity required by a projectile to enable it, without any further source of power, to escape from the gravitational field of the earth or a celestial body. The escape velocity varies with the mass and diameter of the body but not with the mass of the projectile. It is also independent of the angle of launch. For the earth the escape velocity is 11 200 metres per second and for the moon, 2370 metres per second.

Etalon. An interferometer used to study fine spectrum lines based on the interference effect produced by multiple reflection between fixed, parallel, half silvered glass or quartz plates.

Ether (Aether). An imaginary fluid that permeate all space and introduced as a medium through which light as electromagnetic waves can propagate.

Ettingshausen Effect. If a current passed down a conductor which is placed in a magnetic field that is orthogonal to the direction of current flow, then a temeperature gradients is established at right angles to both the current and the magnetic field, and is given by

$$\frac{dT}{dy} = P.B_z I_x$$

where P is the Ettingshausen coefficent, B_z the magnetic flux density and I_x the current.

Euler Force. The critical force required to cause the buckling of a beam supported at both ends. It is given by

$$\frac{\pi^2 EI}{l^2}$$

where E is *Young's Modulus* for the beam, I the moment of inertia about a horizontal axis through the centre of mass and I the length. The beam may not necessarily completely collapse.

Evaporation. The convertion of a liquid into a vapour at temperatures below its boiling point. As it is the most energetic atoms and molecules that escape from the surface in evaporation, the average energy of those that remain is reduced and consequently the liquid is cooled.

Evershed Effect. Spectrographic evidence that the general motion of gases in the penumbral regions of sunspots is radially outwards. The radial velocities are greatest near the solar limb and least at the centre of the disc.

Exchange Force. A force existing between two particles which results due to the exchange of another particle between them e.g. The nuclear force between two nucleons, results from the exchange of mesons between them.

Excitation. In physics, the raising of a system from its lowest energy level (the ground state) to a higher energy level (the excited state). Usually confined to atoms, molecules, ions and nuclei, it is most frequently caused by the absorption of a *photon*. In an atom, ion, or nucleus the photon is absorbed by an electron or *nucleon* causing it to move to a higher energy level. Molecules can be also excited into higher states of rotational or vibrational energy.

Excitation Energy. The energy required to change the energy level of an atom or molecule to a higher energy level. It is equal to the difference in energy of the levels and is the difference in energy between the ground state of the atom and a specified excited state.

Exciton. An electron in combination with a hole in a crystalline solid.

The electron gains sufficient energy to be in an excited state and is bound by electrostatic attraction to the positive hole. The excition may migrate through the solid and eventually the hole and electron recombine with emission of a photon.

Exoergic Process. Refers to a nuclear reaction in which energy is given out.

Expanding Universe. The theory that the universe is expanding was first confirmed by Edwin *Hubble* in 1929 following observations that the light from distant galaxies is subject to a *redshift,* which arises from the recession of the galaxies from us (and from each other). The expansion can be explained by the *big-bang* theory.

Expansion. The general increase in dimensions of a substance with increasing temperature. In solids it is caused by the greater vibrational energy of the atoms leading to increased interatomic distances. In liquids and gases the expansion is caused by the greater velocities of the atoms or molecules. The expansion of an ideal gas is described by *Charles's law.* The *expansivity* or *coefficient of linear expansion* of a susbstance is its increase in length per unit length caused by a rise in temperature of 1° C.

Exposure Meter. A device for measuring the intensity of light falling on a photographic *camera,* used to determine the film exposure time and lens *f-number* needed to suit the lighting conditions. It often consists of a *photocell* that is directed at the subject from the camera, registering light as a reading on an electric current meter. In some cameras this current is used to control the aperture of shutter speed automatically.

Expansivity. A measure of the tendency of a substance to expand on increasing its temperature. The expansivity of a rod of length l_o, α_y is defined by the equation $\alpha = \frac{l - l_0}{l_0 t}$ where *l* is its length at temperature *t*. It is also called coefficient of linear expansion or *linear expansivity*. Similarly, when we consider change in area, the *superficial expansivity* β is defined as $\beta = \frac{A - A_0}{A_0 t}$ and the volume expansivity γ is defined as $\gamma = \frac{V - V_0}{V_0 t}$ α, β and γ are related as

$$\alpha = \frac{1}{2}\beta = \frac{1}{3}\gamma.$$

Extrinsic Semiconductor. A semiconductor whose current carrier density is mainly due to the presence of impurities or other imperfections, as opposed to an intrinsic semiconductor whose electrical properties are characteristic of the ideal crystal.

Eye Piece. A magnifier, generally a combination of lense that give a magnified, aberration free image in a telescope or a microscope. Normally fitted with crosswires which help in the location of the image. Generally consists of two lenses—the lens, towards the eye being called the eye lens, while the other is known as the field lens.

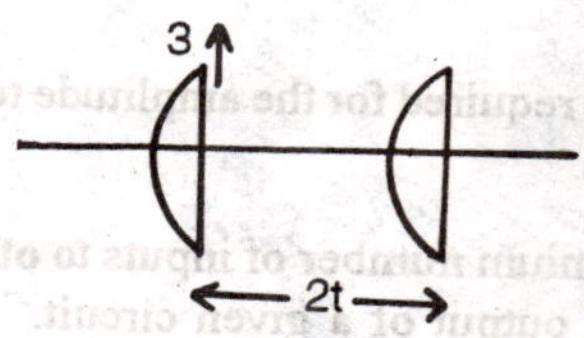

F

Fajans' Precipitation Rule. A radio-element will be co-precipitated with another substance if the conditions of its precipitation are such that the element would form a sparingly soluble compound if it were present in weighable quantities.

There are many exceptions to this rule, but, as modified by Hahn, the following is strictly true:

However, greately diluted, an element will be carried down by a crystalline precipitate, provided that it can be built into the crystal lattice of the precipitate. (Also known as *Paneth's Absorption Rule*).

Fall Time. The time required for the amplitude to fall from 90% to 10% of its peak value.

Fan-out. The maximium number of inputs to other circuits which can be driven by the output of a given circuit.

Faraday. The quantity of electricity that is required to liberate or deposit 1 gram-equivalent of an ion i.e. 96,490 coulombs.

Farad (F). The *SI* unit of electrical capacitance equal to the capacitance of a capacitor across its plates that, if charged with one Coulomb (C) has a potential difference of 1 volt (V), i.e. $1\ F = 1\ CV^{-1}$. Named after Michael *Faraday*.

Faraday Constant. The quantity of electricity equivalent to 1 mole of electrons, i.e. the product of *Avogadro's* number and the electronic charge. It has the value 96 487 coulombs per mole.

Faraday Dark Space. In a discharge tube, the dark space occurring between the positive column and the negative glow.

As the pressure is gradually reduced in a discharge tube filled with, say, air the discharge along the tube can be separated into the following regions, from anode to cathode.

(1) the pink glow immediately around the anode;

(2) the positive column of luminous gas extending from the anode to the beginning of ;

(3) the Faraday dark space;

(4) the negative glow, a pale violet discharge abruptly terminating at the;

(5) *Hittorf* or *Crookes Dark Space;*

(6) a red or orange glow;

(7) a very thin totally dark space adjacent to the *Cathode* called the *Aston Dark Space.*

Faraday Effect. The rotation of the plane of polarization of plane-polarized light when it travels through certain substances in a direction parallel to the lines of force of an applied magnetic field. The effect occurs in quartz and water and is named after Michael Faraday.

Faraday's Laws of Electrolysis. Two laws formulated by M. Faraday in 1813-14. (1) The mass of a susbstance produced by an electrolytic reaction varies directly with the amount of electricity passed through the cell. (2) The masses of substances produced by a given amount of electricity are proportional to the equivalent masses of the substances.

These empirical laws are now understood to hold simply because electricity is composed of uniform discrete partcles (electrons).

Faraday's Laws (of Electromagnetic Induction). (1) When the flux of magnetic induction through a current is changing, an electromotive force is incluced in the circuit. (2) The magnitude of the electromotive force is porportional to the rate of change of the flux.

Far Point. The farthest point from the eye where an object can be seen clearly by the eye. For any object placed of this point, the image is formed at the retina.

Fast Neutrons. Neutrons produced in nuclear fission they have energies in excess of 0.1 MeV and travel at high speeds. Such neutrons are reduced to lower speed or thermalised by making them to collide with suitable susbstances.

Fast Reactor. A nuclear reactor (see nuclear energy) in which natural uranium enriched with uranium-235 or plutonium-239 is used without a moderator, the chain reaction being sustained by fast neutrons. In these reactors the core is surrounded by a blanket of natural uranium into which neutrons escape. These neutrons collide with U-238 nuclei to form U-239 nuclei, which decay to the fissionable isotope Pu-239. By suitable design, more Pu-239 can be produced in the blanket than is required to enrich the fuel in the core. These reactors are therefore called *breeder reactors* and they are 50 times more economical in uranium usage than *thermal* reactors. Their main disadvantage is that the temperature is so high that a liquid metal (usually sodium) has to be used as coolant: any leakage of sodium could be disastrous. Also, plutonium is both extremely toxic and can be used to make *nuclear* weapons. Fast reactors are in use in the UK, Soviet Union and France. Beacause they are so economical in uranium, they are expected to provide a new generation of reactors by the 2020s.

Fax (Facsimilet Transmission). A method of sending images (text or picutres) by a telecommunications link. Most fax transmissions are via the normal telephone network. The document is scanned opticadly, line by line, to break it into a pattern of dots, which are transmitted as electrical pulses. The receiving device preconstructs the image from this digital information and outputs it on photosensitive paper or, in some machines, xerographically. The basic technique is not new-for years it has been possible to 'wire' photographs—but recent advances in technology have made it cheap and reliable, and it is now in widespread use.

Fechner's Law. The eye can distinguish differences in illumination which are a constant small fraction (approximately 0.01) of the total illumination; i.e., it is the ratio rather than the arithmetic difference which is important.

Feedback. In electronics and communications theory, the process of returning to the input of a device a fraction of the output signal. *Negative feedback,* in which the feedback opposes and therefore reduces the input, is often used in *amplifier* circuits. It compensates for noise and distortion in the output signal, although it also reduces the overall amplification. *Positive feedback* reinforces the input signal. If it becomes too high, the circuit oscillates and the output becomes independent of the input. This is the cause of the singing noise heard in a public-address system when the microphone picks up feedback from the loudspeakers.

The principle of negative feedback also operates in biological systems, especially in controlling some biochemical reactions: the Product of the reaction inhibits further activity of the enzyme catalysing it.

Femto. A prefix denoting a factor of 10^{-15}.

Fermat's Principle. When light travels between two points its path is such that the time taken is a minimum. The principle holds when the light is reflected or refracted between the two points. It is also known as the principle of least time and is named after P. *Fermat.*

Fermi Age Theory. Method of calculating the slowing-down density of neutrons in a nuclear reactor. The age equation:

$$\Delta^2 q - \frac{dq}{d\tau} = 0$$

relates the slowing-down density q to neutron age, τ.

Fermi Constant. An interaction or coupling constant for interactions involving a total of four Fermi-Dirac particles including the decay products. It has a value of approximately 10^{-36}J m^3. Examples of Fermi-Dirac particles that may be involved are the neutron, proton, electon, neutroino and muon.

Fermi-Dirac Distribution Function. The probability that an electron will occupy a particular quantum state when thermal equilibrium is reached.

Fermi-Dirac Statistics. A quantum-statistical method of analysing a system of indistinguishable particles to determine the probability of the energy distribution. Unlike *Bose-Einstein* statistics it assumes that these particles, which are known as fermions, obey the *Pauli* exclusion principle. Named after E. *Fermi* and P. A. M. *Dirac.*

Fermi Gas Model. A model of the nucleus wherein the neutrons and protons are considered as independent particles obeying Fermi-Dirac statistics but confined within a cube of volume equal to that of the nucleus.

Fermion. Any elementary particle that obeys *Fermi-Dirac* statistic. These particles, which have half-integral spin, include *leptons* and *baryons*.

Fermi Level. Refers to the highest occupied energy level at absolute zero. It lies in the conduction band in conductors, in valance band in insulator and in the forbedden gap in semiconductors.

Fermi Selection Rules (for Beta Decay). Fermi suggested that if the total angular momentum of a nucleus is I, then, for the emission of a beta particle, $\Delta I = 0$ (the emitted beta particle and neutrino have opposite spins) and there is no change of parity (i.e. no change of right-left symmetry). It was later found that the beta particle and the neutrino could have parallel spins, giving rise to the *Gamow-Tweller Selection Rules:*

$\Delta I = \pm 1$ or (units of $h/2\pi$) with no change of parity, but with $I_i = 0$ to $I_f = 0$ not allowed where subscript *i* refers to the initial momentum and f the final momentum. Beta decay is a so-called 'weak interaction' and it is now known that parity is not always conserved in weak interactions.

Fermi's 'Golden' Rule. If a quantum system has two energy levels and is subject to an external perturbation, the transition probability between the lower state and the upper state at resonance is found to be proportion to the square of the matrix element of the perturbing term and to the density of final states available: i.e.

$$\rho_{kn} = \frac{\pi |H'_{kn}|^2}{h} \rho(E_k)$$

where *k* and *n* are the final and initial states, H_{kn} is a matrix element obtained by integration over spatial coordinates and $\rho(E_k)$ is the density of final states. For resonance, the applied frequency must equal to $\omega_{kn} = (E_k - E_n)\, h$ where E_k and E_n are the energies of states *k* and *n* respectively.

Fermi Surface. In a metal the Fermi energy is a function of magnitude and direction of the momentum of the electrons. Hence, the lengths of momentum vectors for electrons at the Fermi energy define a three-dimensional Fermi surface. If the energy of the electrons is proportional to the square of momentum, then the Fermi surface is sphere. The velocity corresponding to the Fermi energy is called the *Fermi Velocity* and the wavelength calculated from the wave-vector is the *Fermi wavelength*.

Fermi Temperature. The Fermi temperature of an electron gas is defined by E_f/K_B, where E_F is the Fermi energy. It is not the actual temperature of the gas, but corresponds to the temperature at which the electron gas would become classical in its statistical behaviour.

Fertile Material. Substances that can be converted to fissile material by absorption of neutrons.

Feynman Diagram. A method used in calculations involving interactions between many bodies (for instance nuclear particles) to show how a propagating particle starting at point r_1 at time t_1 moves for r_2 at time t_2 interacting during its passage with one or more other particles in the system. The propagation is divided into the probabilities for different types of interaction using the notation.

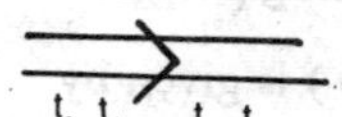

Total motion of a free particle from r_1 to r_2 in time t_1 to t_2.

Probability of free motion from r_1 to r_2.

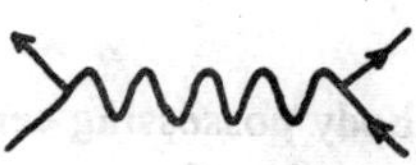

Probability of interaction between a particle at r_1 and a particle at r_2.

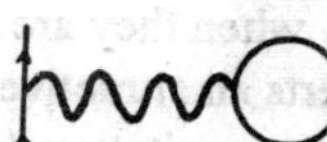

Probability of instantaneous forward scattering with no change of momentum.

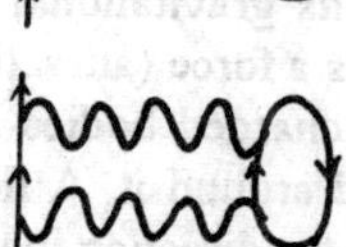

Probability of a chain of collision processes in which momentum at the beginning equals momentum at the end (analogous to the flow of current in an electrical network where 'Flow in' equals 'flow out').

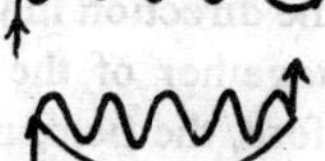

Probability of exchange scattering where the incoming particle is exchanged for the target particle.

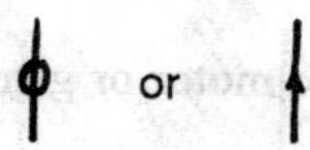

Scattering at 0.

The important feature of a Feynman diagram is how the various lines are connected i.e. the topology of the diagram.

Fibre Optics. The use of flexible glass fibres for transmitting light. Each fibre, which may be used singly or in bunches, is usually less than a millimetre thick and has a high refractive index. The light inside the fibre is totally internally reflected and travels through the fibre with little loss of intensity. The fibres are highly polished and coated with a susbstance of lower refractive index further to reduce dispersion. Glass fibres are used for examining otherwise inaccessible places, for example, in medical diagnosis and in specialized industrial processes.

Fick's Law. (1) The flux of particles J diffusing across a plane in unit time is given by

$$J = -D\frac{dc}{dx}$$

where D is the diffusion coefficient (the flux per unit concentration gradient) and *dc/dx* is the concentraction gradient.

(2) The change in concentration of the diffusing particles with time ($\delta c/\delta x$) for unsteady conditions ($\delta c/\delta x \neq 0$) is given by.

$$\frac{\delta c}{\delta \tau} = D\frac{\delta^2 c}{\delta x^2}$$

Field . In physics, a region of space in which a body possessing certain properties can exert a force on similar bodies, when they are not in contact. For example, a body having mass exerts an attractive force on all other massive bodies as a result of its gravitational field. Similarly, an electrically charged body exerts a force (attractive or repulsive, depending on polarity) on other charged bodies and a magnetized body will have a magnetic field around it. A field is often represented by lines of force to indicate the direction in which the force acts at that point. The closeness together of the lines represents the strength of the force, and therefore, the field, in that area.

Field Coil. The winding (or coil) used in an electric motor or generator that produces the magnetic field when a current is passed through it.

Field Emission. The emission of electrons from the surface of certain materials when they are subjected to a very high electric field. Trypically fields of about 10^{10} volts per metre are required. The effect is also known as cold emission or autoemission. It is utilized in the *field-emission microscope,* in which a magnified image of a surface is obtained by subjecting it to a high electric field and observing the distribution of emitted electrons on a cathode-ray screen.

Field Lens. Refers to the lens farthest from the eye, in the eye-piece system of optical instruments.

Field Strength. (electric, magnetic, etc). See electric field strength, magnetic field.

Filament. A thin wire of tungsten or other metal having hight melting point, used in incandescent electric light bulbs and thermionic valves which is heated by passing an electric current.

Filter. A device or electronic circuit that permits only certain frequencies to pass through and reduces the effect of or stops other frequencies.

Film. A thin flexible strip of cellulose acetate, or similar transparent plastic, coated with a light-sensitive emulsion. A black-and -white photographic emulsion usually consists of gelatin containing tiny suspended crystals of silver halide (usually bromide or chloride). After exposure to light in a *camera* these crystals are easily reduced to metallic silver when treatd with the chemicals in the developer. This produces black deposits of fine particles of metallic silver on the parts of the film upon which the light has fallen, giving a reversed (or negative) image. Fixing of the film consists of bathing it in sodium thiosulpyhate (hypo) or other fixers to render the unchanged silver halides soluble, enabling them to be washed away with water. The sensitivity (speed) of film is usually quoted as an ASA rating, which determines the amount of light required to form a given amount of metallic silver.

Fine Structure. The splitting up of certain spectrum lines on being examined under high resolution. Each line can be resolved into a number of closely spaced lines.

Fissile Material. Substances that are capable of undergoing nuclear fission, generally upon impact with a thermal neutron.

Fission Products. Refers to both the stable and the unstable isotopes produced as a result of nuclear fission.

Fission-track Dating. A method of dating based on the sponteneous nuclear fission of uranium-238 in the sample. The fissions are recorded as tracks, which are then compared to the tracks formed by inducing fission in the uranium-235 present. A comparison of the numbers of tracks is used as a measure of the age of the sample.

Fixed Point. An accurately reproducible equilibrium temperature e.g. the ice point, the steam point etc.

Fizeau Fringes. Interference fringes observed in a wedge-shaped film due to interference between waves reflected from the two surfaces. With an extended white-light source, the fringes coincide at the film and are said to be localized in the plane of the film. With a point source the fringes are no longer localized in this way.

Flash Barrier. A structure of fire-proof material so as to minimize the formation of electric arc between conductors or to minimize the damage caused by such an arc in an electric machine.

Fleming's Rules. A mnemonic, invented by Sir John Ambrose Fleming, for the relationship between the directions of motion, magnetic field, and electric current in *electric motors* and *generators*. The thumb represents motion; the first (index) finger represents the field; and the second finger represents the current. If the right hand is held with the thumb and first two fingers straight and mutually at right angles, the directions in a generator are indicated. The left hand indicates the directions in a motor.

Flicker Photometer. Photometer used for comparing illuminating powers of different sources of light. In this light falls on two surfaces from the two sources in rapid succession. If the illumination of the two surfaces is unequal, an unpleasant flicker is produced. The distances of the two sources from the respective surfaces are adjsuted till the flicker disappears. The photometers can be used even if the sources emit lights of different wavelengthes.

Floatation Law of (Principle of Floatation). A body floating in a fluid displaces a mass of the fluid that is equal to its own mass.

Fluidics. The use of jets of fluid in a circuit to carry out electronic functions. Fluidic circuits can resist much higher temperatures than electronic circuits and are also unaffected by ionizing radiation and magnetic fields. They therefore have uses in nuclear reactors and spacecraft. They are also used as delay lines since they respond much more slowly than electronic circuits.

Fluidization. The process of supporting very fine solid particles in a stream of gas so that the combination of solid and fluid behaves like a liquid. The process is used in transporting coal dust and in the cleaning of the catalyst in calystic cracking in oil refining.

Fluid Mechanics. The study of the mechanical properties of fluids. *Hydrostatics* is concerned with the study of fluids at rest and hydrodynamics (or fluid dynamics) with fluids that are flowing. Hydraulics deals with the practical applications of these sciences. Two important aspects of hydrodynamics are the conservation of energy in fluid flow and the distinction between streamline and turbulent flow.

Fluoroscence. The luminescence produced when certain susbstances absorb radiation of one wave length and emit light of another wave length. The emission of light stops as soon as the incident radiation is cut off unlike phosphoroscence in which the emission continues even after the incident radiation is switched off.

Fluoroscent Lamp. A glass Tube coated inside with a fluoroscent substance, that act as a light source. The tube contains mercury vapour and contains two electrodes. When a voltage is applied between the electodes, electrons flow through the tube and collide with the mercury atoms that emit ultraviolet radiation. This falls on the fluoroscent walls and visible light is emitted by the fluoroscent material.

Flux (Physics). The net amount of a directional quantity passing through a surface area at right angles to the surface. The concept of flux is used to describe phenomena that involve forces or the flow of energy, such as electric flux, magnetic flux and luminous flux.

Flux Density, Magnetic. Refers to the number of magnetic lines of force per unit area at a point.

Fluxmeter. An instrument used to measure magnetic flux. It consists of a coil connected to galvanometer. When the coil is placed in a magnetic field (flux) and withdrawn suddenly the flux linked with the coil changes. It produces a current in the galvanometer. The galvanometer deflection is a measure of the magnetic flux.

Flywheel. A large heavy wheel attached to the driving shaft of a motor to act as an energy store and to iron out fluctuations in the speed of the machine. The energy stored depends on the speed of rotation and the weight distribution of the wheel. They are used in *internal*-combustion engines, power presses, etc.

F-number. The ratio of the focal length of a camera lens to the diameter of the shutter opening (aperture). For example, f-8 means that the focal length is eight times the aperture. The smaller the f-number, the greater the illumination of the film.

Focal Distance (Focal Length). When a parallel beam of light is allowed to fall on a lens or mirror, it converges to a point, or appears to diverge from a point on its principal axis. The distance between this point, known as focus, and the optical centre of the lens or mirror is called its focal distance or focal length.

Focal Plane. Refers to a plane perpendicular to the principal axis of a lens, or morror and centered at its focus.

Focus. (1) The point on the principal axis at which a beam of converging rays meet, or from which a beam of diverging rays seem to originate from. (2) Principal focus. Refers to a point on the principal axis where a beam of light parallel to the axis converges or appears to diverge from after reflection or refraction.

Forbidden Transition. A transition of an atom or molecule between energy levels involving a change of quantum number which is not allowed by the selection rules.

Force. The agency that changes either the speed or the direction of motion of a body (symbol: F). It is a *vector* quantity defined as the product of the mass of the body and the acceleration produced on it. Force is measured in newtons. *See also* centripetal force; Coriolis force.

Forced Oscillation. Oscillation of a system or object induced by an external periodic force. It has the same frequency as the frequency of the external force.

Formula. A statement of a relationship expressed in terms of symbols of quantities involved. e.g., the formula for the time period T of a simple pendulum of length l is

$$T = 2\pi \frac{l}{g}.$$

Forward Bias or Voltage. A voltage applied to a circuit in such a direction so as to produce the large current.

Foucault Current. A current included in the interior of conductors by variations of magnetic flux.

Foucault Pendulum. A very long pendulum with a heavy bob, capable of oscillating for a long period. It demonstrated the earth's rotation since, as it swings, its plane of oscillation slowly rotates. At the poles it would rotate through 360^0 in 24 hours; at a latitude λ, the number of hours for a complete rotation is given by $24/\sin\lambda$. It was invented by Jean *Foucault* and first demonstrated in 1851 in Paris using a 28-kg lead ball suspended from a wire 67m long.

Franck-Condon Principle. Electronic transitions in molecules occur in times which are very short in comparison with the period of vibration of the nuclei. Hence, in an energy level diagram, the most probable electronic transitions occur vertically.

Franklin Equation. This relates the energy level of sound in a room as a function of time after the source has been cut off.

$$E = E_0 \exp(-c\alpha St/4V)$$

Where E and E_0 are the final and initial sound levels, c is the velocity of sound. S the surface area, V the volume of the room, t the time and α is the mean sound absorption coefficient.

Franz-Keldysh Effect. A shift to longer wavelengths, in the absorption edge of semiconductors, produced by a strong electric field.

Thus optical transmission and reflectance can be modulated by the application of an electric field.

Fraunhofer Diffraction. Diffraction phenomena may be divided into two classes:

(1) When the light source and the screen upon which observations of the pattern are taken at an infinite distance from the aperture, the diffraction is known as Fraunhofer diffraction.

(2) When either the source or the screen, or both, are at a finite distance, then it is known as *Fresnel Diffraction.*

Fraunhofer Lines. Dark absorption lines found in the spectrum of the sun as observed from the surface of the eaarth and designated by letters A, B, C,... starting at the red end. It is customary to use light of the corresponding frequencies when specifying indices of refraction for optical materials.

Free Energy. (Gibb's Function, G). The energy that is absorbed or evolved during a reversible thermodynamic process. Mathematically G = H - T S, where H is the enthalpy (heat content), T the temperature and S, the entropy of the system.

Frenkel Defect. A defect in a crystal lattice formed because of an atom of ion being removed from its normal position in the lattice can taking up an interstitial position.

Frenkel Excition. An exciton occurs in a crystalline solid and may be considered as a conduction-band electron and a valence-band hole bound together (although with finite separation) moving through the crystal. A Frenkel exciton is a tightly bound electron-hole pair with small separation of the electron and hole. A *Wannier* (or *Mott*) *Exciton* is a weakly bound pair with large separation.

Frequency Divider. An electrical apparauts of curcuit the output frequency of which is an exact integral sub-multiple of the frequency of the input. A common form employs a vibrator.

Frequency Meter. An instrument for measuring the frequency of an alternating current. A common type consists of a cavity reasonator that measures the frequency of an electromagnetic wave.

Frequency Modulation. Modulation of the frequency of a carrier wave in accordance with the audio or visual signal to be transmitted. It is superior to amplitude modulation as a broader frequency range can be transmitted with reduced noise and interference.

Frequency of a Vibrating String. The fundamental frequency, f, of a stretched string of length l under tension T, is given by

$$f = \frac{1}{2l} \frac{T}{m}$$

where m is the mass per unit length of the wire.

Fresnel. A unit of frequency equal 10^{-12} hz.

Fresnel's Biprism. A glass prism with an obtuse angle (nearly 178^0) that produces two coherent virtual sources and interference pattern

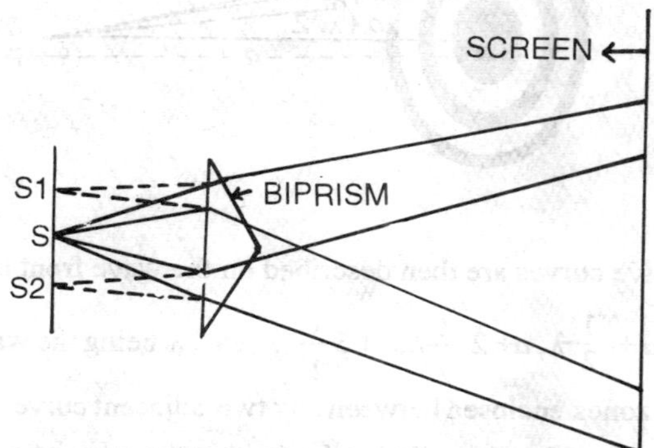

when monochromatic light is allowed to pass through it. The fringe width depends on the wave length of light used and its measurement enables one to calculate the wave length accurately.

Fresnel-Arago Laws (of Polarized Interference). (1) Two rays polarized at right angles do not interfere. (2) Two rays polarized at right angles obtained from the same beam of polarized light will interfere in the same manner as ordinary light only when brought into the same plane. (3) Two rays, polarized at right angles and obtained from perpendicularly polarized components of unpolarized light, never interfere even after rotation of their planes of polarization.

Fresnel Diffraction. Refers to diffraction observed when either the source or screen or both are at a finite distance from the aperture or obstacle.

Fresnel Lens. A lens having a surface made up of a large number of smaller lenses to produce a short focal length lens.

Fresnel Zone. In considering the effect of a wave front at any point P, it is convenient to separate the wave front into regions formed first by constructing a tangent sphere to the wave front with centre at P and radius, say, a.

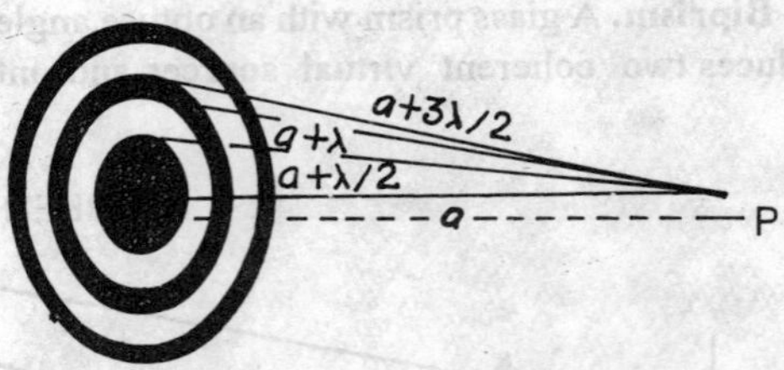

Successive curves are then described on the wave front by spheres of radii $a+\frac{1}{2}\lambda$, $a+2\frac{1}{2}\lambda$, $a+3\frac{1}{2}\lambda$, etc., λ being the wavelength, and the zones enclosed between any two adjacent curves are called Fresnel zones. It can easily be shown that the areas of these zones are equal.

Friction. A force exerted at the boundary between two solids or fluids that retards motion between them. In solid friction a distinction is drawn between sliding friction and rolling friction. Sliding friction is further divided into dynamic friction, defined as the minimum force needed to keep a body sliding, and static friction, defined as the minimum force needed to move a stationary body. The latter is slightly greater than the former. In rolling friction the force of resistance is less than in sliding friction as the rolling body moves up the side of a depression made in the stationary body. This accounts for the effectiveness of wheels and ball bearings. The coefficient of friction is defined as the ratio of the frictional force to the perpendicular reaction between the surfaces. Friction is caused primarily by the two surfaces interlocking at the microscopic level. It is reduced by the use of lubricants, such as grease or graphite. Where it is sometimes desirable to increae friction between surfaces, as in brake linings, clutch plates, and shoe soles, rough hard-wearing materials are used.

Friedel's Law. Every crystal diffracts X-rays as if a centre of symmetry were present. Thus the diffraction symmetry of a crystal is its point group symmetry plus a centre of symmetry, and corresponds to any one of eleven kinds of centrosymmetry (the *Laue Symmetry Groups* i.e., the intensity of reflection from opposite sides of the same set of crystal planes is the same. The law can fail in polar crystal for radiation which is close in wavelength to an absorption edge.

Fringes. Alternate dark and light region obtained by interference or diffraction of light.

Fuel Calorific Value of. The amount of energy produced when a unit mass of the fuel is completley burned.

Fuel Injection. The pumping of fuel in the form of a spray directly into the cylinders of an **internal-combustion* engine. This is necessary in the case of continuous-combustion engines fuel injection is also used; in petrol engines it is increasingly used to replace the normal carburettor as it gives a more even fuel distribution in the combustion chamber. Car models using fuel injection usually have the initial "i" after their model number.

Fuel Wave Rectifier. A rectifier that converts the negative half wave of an alternating current into a positive half wave. So that both halves of the swing deliver a undirectional current.

Full Load. The maximum output of an electrical machine under certain specified condition.

Function Generator. A signal generator that produces specific waveforms that may be used for test purposes, over a wide range of frequencies.

Fundamental. Generally that component of a complex vibration which describes the patch of a note. In a given note it is generally the tone with the lowest frequency.

Fundamental Constants (Universal Constants). Some quantities that do not change as the environment changes e.g. Planck's constant, velocity of light in vacuum.

Fundamental Units. Refers to the units of length, mass and time in terms of which the units of all other mechanical quantities can be defined. In S.I. system the units of mass length, and time are metre, kilogram and second respectively.

Furlong. A unit of length traditionally based on the length of a furrow. It is equal to 220 yards (1/8 mile) and is still used in horseracing.

Fuse Electrical. A device that prevents a high electric current from passing through a circuit. It consists of a piece of wire of metal of low melting point. A large current will increase the temperature of the fuse wire sufficiently to melt it and thus break the circuit.

Fusion. (1) *Syn.* melting. The change of the state of a substance from solid to liquid that occurs at a definite temperature (melting point) at a given pressure. (2) nuclear fusion.

Fusion Reactor. *Syn.* thermonuclear reactor. A device wherein nuclear fusion takes place and where there is a net evolution of usable energy.

G

Gallon. An Imperial unit of volume equal to the volume occupied by ten pounds of distilled water under defined conditions. The US gallon is equal to 0.832 68 British gallons.

Galvani, Luigi (1737-98). Italian physician, who poineered research into the electrical properties of living things. He observed how frog muscles twitched when they were touched by metal contacts but he wrongly attributed this to innate "animal electricity" (the current was actually produced by the metal contacts). The *galvanometer* was named after him.

Galvanometer. An instrument for measuring small electric currents. The moving-coil galvanometer consists of a coil of wire suspended in a magnetic field. A current passing through the coil causes it to rotate until balanced by the opposing torsion in the suspending thread. The angle of rotation is used to measure the current. In the moving-magnet instrument the mangent is suspended in the earth's magnetic field and deflection is caused by a current passing through the surrounding coil.

Gamma Rays (Gamma Radiation). Electromagnetic radiation having very short wave length. They are given out during decay of radioactive nucli and when a nuclide in an excited state makes transition to its ground *state*.

Gamow Factor. An exponential factor e^{-G} which determines the yield of nuclear reactions. At low energies, G has the approximate form (in C SI units) of

$$\frac{zZe^2}{2\varepsilon_0 h\,v}$$

for a particle of charge *ze* approaching a nucleaus of charge *Ze* at a velocity *v*.

Gamut. The interval between a note and its octave is generally divided into seven smaller intervals. The eight notes so formed constitue a musical scale or a gamut.

Gantmakher Effect. If a very thin plate of metal is placed in a magnetic field with the field parallel to the surface of the plate, and microwave radiation polarized perpendicular to the direction of the magnetic field is incident orthogonally on to the surface of the plate, then there will be resonant transmission through the plate for certain thickness or for certain values of the magnetic field. This resonant transmission corresponds to the thickness of the metal sheet equalling either the extremal orbital diameter of the electrons, as allowed from the shape of the *Fermi Surface* of the metal, or integral multiples of the diameter.

Gas Thermometer. A thermometer that is used to measure temperature by observing the variation in the properties of a gas as a function of temeperature. The gas may be kept at constant volume or constant pressure corresponding to which we have the constant volume gas thermometer or constant pressure gas thermometer.

Gas Turbine. A form of *internal-combustion* engine consisting of a *turbine* in which the power to drive the blades is provided by hot gas. The gas turbine is a flexible engine with many applications, the most useful being in aviation. It consists of a *compressor,* a combustion chamber, and the turbine. Atmospheric air is fed under pressure from the compressor to the combustion chambers, where a fuel, such as natural gas, paraffin, or oil is burnt; the hot gases then drive the turbine, which in turn drives the compressor. Power is supplied either in the form of thrust from a jet or rotation of the turbine shaft. Heat engine.

Gauss (G). The unit of magnetic flux density in the *c.g.s.* system equal to a flux density of one maxwell per square centimetre. Named after Karl Friedrich *Gauss*.

Gauss' Theorem. According to this the total electric flux over a closed surface is equal to the $\frac{1}{\varepsilon_0}$ times the total charge contained within the surface, where ε_0 is the permitivity of the free space.

Geodesic. Refers to the shortest distance between two points on a spherical surface.

Geilger, Hans (1882-1945). German physicist, who in 1913 invented the *Geiger* counter for detecting ionizing radiation in connection with his work on cosmic rays. During World War II he participated in Germany's unsuccessful attempt to build an atomic bomb.

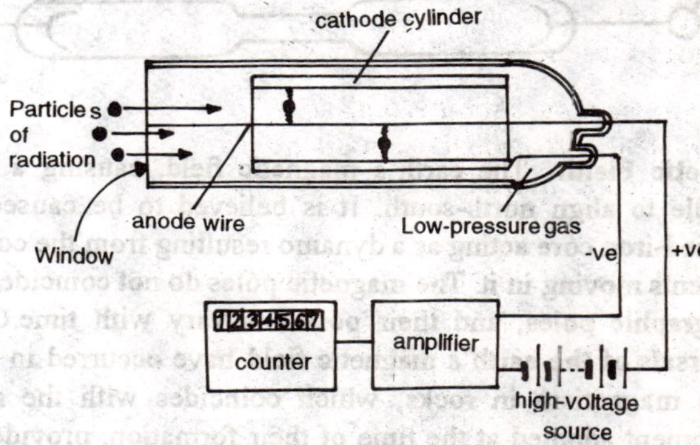

Geiger Counter. The radiation entering the tube causes an electrical discharge through the gas between the anode and cathode, sending an electrical pulse to the counter.

Gelger Counter. A device that detects and counts ionizing radiation and particles. Essentially it consists of a meteal cylinder containing low-pressure gas and a wire anode running along its central axix. The anode is held at a potential difference just less than that required to produced a discharge in the gas. Ionizing particles passing into the tube through a window at one end induce discharges, which can be counted by a suitable circuit. Named after Hans *Geiger*.

Geiger-Nuttal Rule. An empirical relationship between the radioactive constant λ and the range R of α-particles emitted by a radioactive element.

$$\log \lambda = b + c \log R$$

where b and c are constants, b differs for each of the three radioactive series and c is practically the same for each series.

Geissier's Tube. A vacuum tube evacuated to a pressure of 1 mm of mercury. The discharges concentrated in a narrow region to make it brighter. The tube is used to study the spectra of gases.

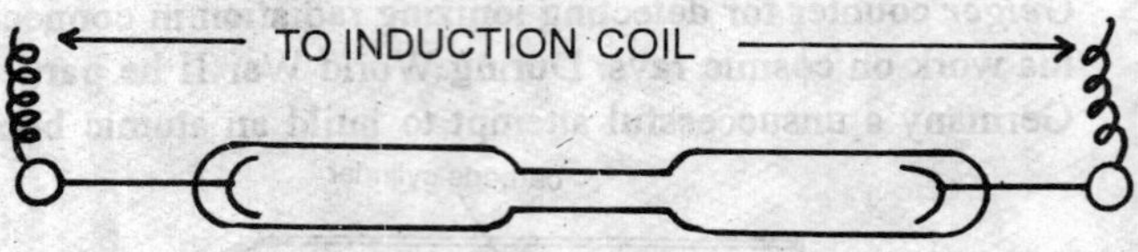

Geomagnetic Field. The earth's magnetic field, causing a compass needle to align north-south. It is believed to be caused by the liquied-iron core acting as a dynamo resulting from the convection currents moving in it. The magnetic poles do not coincide with the geographic poles, and their positions vary with time. Complete reversals of the earth's magnetic field have occurred in the past; relic magnetism in rocks, which coincides with the magnetic alignment adopted at the time of their formation, provides strong evidence for the theory of sea-floor spreading and continental drift. The three *magnetic elements* of the earth's field are the *magnetic* clip, the *magnetic* declination, and the horizontal field strength, which together completely define the earth's field at any point on its surface.

Geometrical Resolving Power. Refers to the smallest angle subtended at the instrument, whose resolving power is being defined by two objects which can be seen as just separate and distinguishable.

Generator. A mechine that produces electrical energy from mechanical energy. It is made up of two sets coil windings, one of these is fixed and the other, known as rotor, rotates. The relative motion between the two sets of coils induces an electric current in them.

Geoid. The geoid is a surface in the earth's gravitational field and is approximately an oblate spheroid in shape.

Geometrical Optics. Study of optical phenomena that can be discussed in terms of light rays and in which the wave nature of light does not play an important part. Formation of images by mirrors and a lenses is part of the geometrical optics.

Geophysics. The study of the physical forces acting on, and particularly within, the earth. Sophisticated equipment is used to sutdy the properties, structure, composition, and evolution of the earth. Important branches of geophysics are seismology, geomagnetism, vulcanology, natural radioactivity, and the earth's rotation and gravitational field. Much geophysical data has been collected in recent years as a result of geophysical prospecting, particularly for petroleum.

Getter. A material having a strong chemical affinity for other materials. Such materials may be used to remove unwanted atoms or molecules from an environment; for example barium may be used in a sealed vacuum system to remove residual gases or phosphorus may be introduced into oxide layers on silicon to remove mobile impurities, such as sodium.

GeV (1 giga electron Volt). A unit of energy, 1 GeV= 10^9 eV Also written as 1 BeV (called 1 billion electron volt).

Geysers. Jets of hot water and steam issuing intermittently from holes in the earth's crust, some reaching heights of up to 70n. Geysers are found in volcanically active or recently active regions, for example, in Iceland. They occur when water from deep within the crust becomes superheated and suddenly boils, gushing up to the surface like a fountain. Cones of sinter (deposits of silica) frequently build up around the vents of geysers.

Ghost. Image formed by reflection from the front surface of glass. Also, a false line appearing in the spectrum produced by a diffraction grating due to imperfections in the grating. *Ghost television images* apppear on the TV screen if the TV signals reach the aerial after reflection from a nearby high building.

Gibbs-Duhem Equation. A relation between the chemical potential and concentration of species in a mixture at constant temeperature and presure. If there are n_i moles of species i present having chemical potential μ_i, then

$$\sum_1 n_i\, d_i \mu_i = \sum_1 x_i\, d\mu_i = 0$$

where $x_i = n_i / \Sigma n_i$ is the mole fraction and $d\mu_i$ is the change in chemical potential for a small change in the number of moles dn_i.

Gibb's Function (or Free Energy). The Gibb's function, or free energy, G, is given by

$$G = H - TS$$

where H is the enthalpy and S is the entropy.

Gibbs-Helmholtz Equation. This equation was deduced independently by J.W. Gibbs in 1875 and H von Helmholtz in 1882. It may be expressed in a number of ways. If G is the change in free energy during a reaction and ΔH the increase in heat content, then,

$$\Delta G - \Delta H = T\left[\frac{\Delta \delta G}{\delta T}\right]_P$$

Guibbs' Paradox. One might expect that, as two diffusing gases become more and more alike, the change in entropy due to diffusion should get smaller and smaller, approaching zero as the gases become identical. The fact that this is not the case is known as the Gibb's Paradox. Bridgman explained it in the following fashion: in principle at least it is possible to distinguish the dissimilar gases by a series of experimental operations. In the limit, however, when the gases become identical there is a discontinuity in the operations in that no instrumental operation exits by which the gases may be distinguished. Hence a discontinuity in a function such as change in entropy is to be expected.

Gibbs' Phenomenon. In considenring the transient response of an electrical network to a unit step function, the wider the band of frequencies passed by the network the better will be the response. Gibbs, however, stated that no mattter how high is the cut-off frequency of the network, the maximum amplitude of the trailing overshoot of the leading edge never becomes less than 0.175. More generally, a *Fourier Series* expansion of a function round a discontinuity results in an overshoot, even if an infinite number of terms is taken.

Gilbert. Symbol: Gb. The CGS-electromagnetic unit of magnetomotive force or magnetic potential. One turn of wire carrying a current of 1 ampere produces a magnetomotive force of 4π/10 gilberts.

Giorgi System. A system of measurement in which the units are the metre, the kilogram, the second and the ampere.

Giorgi Units. Units based on the metre, kilogramme, and second as fundamental units along with one electrical unit of practical size. Now superseded by SI units.

Glide Plane. A plane upon which glide can take place upon application of a suitable shearing stress.

Gilders. Light fixed-wing engineless aircreaft, sometimes called sail-planes. They are launched into the air by a winch or catapult or by being towed by a car or powered aircraft. Once airborne a glider slowly loses height unless it is lifted by a rising air curent created by warm air rising from the ground (a thermal), a ground contour, or a thunderstorm. The height record is 14.1 km (P.F.Bikle: 1961) and the distance recor 1460.8 km (H.W. Grosse; 1972).

Pioneered by Otto **Lilienthal* in the USA, gliders were used by the **Wright* brothers in designing their powered aircraft. Gliders towed by aircraft were used in World War II to carry men and equipment, notably by the British in the ill-fated Arnhem expedition. Gliding is a popular sport, controlled in the UK by the British Gliding Association (1924).

Glow Discharge. Generally an electric discharge through a gas, at a low pressure, in which the gas becomes luminous.

Gluon. Hypothetical particle exchanged between quarks that bind them together.

Gold Leaf Electroscope. An instrument used for detecting the presence of an electric discharge. It consists of two reactangular leaves of gold foil attached to a conducting rod of metal held by an insulating plug. When the rod and leaves acquire (the same) electric charge, the gold-leaves expand because of the mutual repulsion of similar charges.

Goldschmidt's Law. After consideration of the crystal structures of a large number of inorganic compounds, V. Goldschmidt postulated in 1929 the fundamental law of crystal chemistry.

The structure of a crystal is determmined by the ratio of the numbers, the ratio of sizes and the properties of polarization of its structural units.

Gong. A percussion instrument of indefinite pitch. The orchestral gong (or *tam-tam*) is a disc of metal about 1 m (3 ft) in diameter with a turned-over edge, hanging in a wooden frame.

Gong Chimes. A set of gongs tuned to different pitches, used in many countries of the Far East. The instrument exisis in a number of forms, the most eleaborate of which, found in Burma and Thailand, consists of 15-20 small gongs set in a circular wooden frame. The player kneels inside the frame and plays the gongs with hammers.

Graetz Number. A dimensionless number used in problems of heat transfer and defined as

$$\frac{vp\,\mathrm{AC}p}{\lambda 1}$$

where v is the velocity of flow, p the density, A the surface area, C_p the specific heat at constant pressure, λ the thermal conductivity and 1 a characteristic length.

Grain. A British unit of mass

1 grain = 0.648 gm. = 1/7000 lb.

Gram (g). A unit of mass equal to 1/1000 of a **kilogram*. The gram is itself one of the basic units in the *c.g.s.* system, but in *SI* units the basic unit is the kilogram.

Gravitation. A attractive force that occurs between all bodies that posses mass. It was first described by Sir Isaac *Newton* in a law stating that the force between two bodies is directly proportional to the product of their masses and inversely proportional to the square of the distance between them. The constant of proportionality is called the universal gravitational constant, G, which has the value 6.673 $\times 10^{-11}$ newtron metre squared per kilogram squared. Gravitation is now more accurately described by the general theory of *relativity*. In this theory a mass distorts the *space-time* continuum around it so that the geometry of the space is locally no longer Euclidean. The force of gravity and the *acceleration* of free fall are the result of the attractive force between a body and the earth.

Gravitational Collapse. The sudden collapse of the core of a *star* when thermonuclear fusion eventually ceases. The star's internal gas pressure can no longer support the weight of the star and the initial result may be a *supernova* explosion. The gravitational pull of all the constituents of the star, or its remains, causes it to contract. The extent of the contraction depends on the mass of the object, producing a *white dwarf, neutron star,* or *black hole.*

Gravitational Field. The region around a massive body wherein a force of attraction is felt by other bodies, e.g. a gravitational field exists on and near the surface of the earth and all the bodies on or near the surface of the earth experience a force of attraction towards the centre of the earth.

Gravitational Force. One of the four fundamental forces. The gravitational force (*see* gravitation) influences all matter and acts over an infinite range.

Gravitational Interaction. One of the four kinds of interaction that occur between elementary particles (*see* particle physics) and by far the weakest (about 10^{40} times weaker than the *electromagnetic* interaction). The interaction occurs between all particles with mass and can be explained as the exchnge of *virtual* particles called gravitons. Such particles have not yet been detected.

Gray (Gy). The SI unit of absorbed dose of ionizing radiation equal to the energy in joules absorbed by one kilogram of irradiated material.

Graviton. A hypothetical particle responsible for the effects of gravity.

Gravity. The resultant effect of the gravitational attraction and centrifugal repulsion between any body of matter and a planetary body as given by the weight of the body near the planet; and the acceleration with which the body freely falls towards the planet: commonly denoted by *g*.

Gravity Cell. A primary electric cell wherein two electrolytes are kept apart due to their different densities.

Gravity Wave. A wave occurring in the surface of a liquid, the wave being controlled by gravity and not by the surface tension of the liquid. For example, for shallow sea waves the velocity v depends on the wavelength λ according to $v = (g\lambda)$ ½, where *g* is the acceleration of free fall.

Greninger Net. A net used for interpreting back-reflection Laue photographs. The net can be used for identifying groups of spots which arise as a result of reflections from groups of planes in the same zone, and for measuring angles between spots.

Grid. Refers to an open wire mesh between the anode and cathode of a triode valve to control or modify the current effectively by small voltage changes. Its introduction enables it to be used as an amplifier. Also called control grid.

Grid Bias. Voltage that is applied between the cathode and the control grid of a thermionic valve (like triode); the valve of the grid bias determines the operating conditions of the system.

Ground Waves (*Also called* **Direct Waves**). Electromagnetic radiations that reach the receiving station direct from the transmitting station without suffering any reflection from the ionosphere.

Group Velocity (v_g). Refers to the velocity with which a group of waves travels. It is related to the phase velocity by the equation $v_g = Vp - \frac{\lambda dv_p}{d\lambda}$, equation, where λ is the wave length. If the phase velocity does not depend on the wave length, $\frac{dv_p}{d\lambda}$ and the group and phase velocities are equal.

Gudden-Pohl Effect. The flash of light or luminescence which occurs in certain materials, such as zinc sulphide phosphors, when an electric field is applied or removed while the material is exhibiting phosphorescence (or afterglow). The effect is often called electrophotoluminescence.

Guillemin Effect. A magnetostrictive effect in which, if a bar of ferromagnetic material is eleasticity or permanently bent, it tends to straighten upon magnetization.

Guinier-Preston Zones. In the solid state, a phase of a material may contain micro-constituents which do not show definite boundaries within the phase. These regions are referred to as Guinier-Preston zones.

Gunn Effect. The effect used in solid-state *semiconductor* microwave generators (Gunn diodes). When a sufficiently high steady electric field (typically several thousand volts per centimetre) is applied across a sample of n-type gallium arsenide, microwave frequency current oscillations are set up. Named after J.B.Gunn (1928-).

Gurevich Effect. In an electrical conductor, over which a temperature gradient exists, phonons will themselves carry a thermal current (the lattice heat flow) and if phonon—electron collisions are important—as they are in a pure metal at low temeperatures—the phonons may tend to drag the conduction electrons with them from hot to cold. This effect will produce an additional component of thermoelectric power.

Gyulai-Hartly Effect. An observed increase in electrical conductivity in ionic crystals when compressively deformed. The effect is thought to arise from electrical charges associated with dislocations in the crystals.

Gyrocompass. A non-magnetic compass that uses a gyroscope fitted with a pendulous weight or some equivalent to induce precession due to gravity. It aligns with the true N-S direction.

Gyromagnetic Effects. Refers to the relationship between the mangentization of a body and its rotation.

Gyromagnetic Ratio. It is the ratio of the magnetic moment to the angular momentum of an atom or a nucleus.

Gyroscope. A device consisting of a heavy wheel mounted in a double gimbal so that it is free to rotate about three mutually perpendicular axes. Once spinning, the wheel maintains the same orientation in space, even when the gimbals are turned. This is made use of in the gyrocompass (compass) and the gyrostabilizer.

H

Half-width. The wave length region in which the relative intensity as a function of wave length for an electromagnetic radiation lies between 0.5 and 1.

Hall Effect. The Hall effect is as phenomenon which may be observed when conductors and semiconductors are subjected to an electric and a magnetic field orthogonal to each other. A voltage, the *Hall voltage* V_H appears across the sample in a direction at right angles to both these fields. If the current density arsing from the applied electric field is J_y, the magnetic flux density is B_z and the thickness is Δ_y then a *Hall coefficient* R can be defined:

$$V_H = R j_x B_z \Delta y.$$

It can be shown for the case of electrons in a metal that the Hall coefficient is related to the carrier density n by the relationship

$$R = -\frac{1}{ne}.$$

For electrons in a pure semiconductor, because of a different velocity distribution, this relationship becomes.

$$R = -\frac{3\pi}{8}\frac{1}{ne}.$$

For holes the sign is reversed, and hence the measurement of Hall voltage is useful practically to distinguish between n- and p-type materials. If both carriers are present as in an intrinsic semiconductor, the relationship is more complicated

$$R = -\frac{3\pi}{8e}\frac{n\mu_n^2 - p\mu_p^2}{(n\mu_n + p\mu_p)^2}$$

where μ_n and μ_p are the Hall mobilities of electrons and holes respectively and *n* and *p* are their carrier densities.

The angle between the direction of current flow and the total electric field (i.e., the vector sum of the applied field and the Hall field) is called the *Hall angle.*

Hall Mobility. Symbol: μ_H. The product of the Hall coefficient R_H and the electrical conductivity *k.*

Hallwach's Effect. A negativley charged body in a vacuum is discharged by exposing it to ultraviolet light.

Hamiltonian Function. *Syn.* Hamiltonian. Symbol: H. A function which expresses the energy of a system in terms of momenta, *p,* and the positional coordinates, *q*. It is *much* used in wave mechanics.

Harmonica (1) The mouth organ : the smallest member of the reed-organ family; its invention is attributed to Sir Charles *Wheatstone* in 1829. Notes and chords are obtained by blowing or sucking rows of parallel reeds. It is used mainly in light music, although Vaughan Williams and *Milhaud* have written workers for the intrument's main eponent, larry *Adler.* (2) The glass harmonica: an obsolete instrument consisting of tuned glasses rubbed with a damp finger.

Harmonic Analysis. A procedure, developed in 1822 by Joseph *Fourier,* by which complicated periodic functions, such as sound waves, can be written as the sum of a number of simple wave functions known as a Fourier series. One wave functions, called the fundamental, has the same frequency as the original function. Those that have frequencies that are integral multiples of the fundamental are called harmonics.

Hartmann Test. To determine the aberration in a lens, a disphragm containing a number of small apertures is placed before the lens and the course of the rays is plotted by photographing the pencils of light in planes on either side of the focus.

Hartree-Fock Technique. A method for calculating the effect of interactions between many bodies. It takes into account the exchange interaction between the bodies, although calculation of this interaction is laborious and usually requires approximations. The method has extensive application, particularly in solid-state and nuclear physics.

Hartree System of Units. A system of units in which the values of *h.*, the quantum unit of angular momentum, *e*, the charge on the electron in electrostatic units, and *m*, the mass of the electron are set to unity. In this system the unit of length is h^2/me^2, the radius of the first Bohr orbit for an electron moving non-relativistically around a proton of infinite mass, and the unit of time is h^3/me^4, the time taken for an electron to move in a Bohr orbit through 1 radian.

Head of Liquid. Difference of level between the free surfaces of a liquid and that of the delivery pipe through which the liquid is supplied. Measured in metres.

Heat. The form of energy that is transferred from one body or region to another at a lower temperature. The amount of heat gained or lost by a body is equal to the product of its **heat* capacity and the temperature through which it rises or falls. Heat is transferred by conduction, convection, or radiation. The internal energy of a body, the *total kinetic* energy of its components atoms and molecules, is sometimes called its heat, although this can lead to confusion. Heat is measured in joules, but older units, such as calories and British thermal units, are still sometimes used.

Heat Calorific Theory of. The earliest theory of heat according to which heat is an elastic imponderable yield called calorie. It is assumed to pervade all substances and is considered as indestructible and uncreatable. The theory can explain satisfactorily some thermal processes like the flow of heat from a hot body to a cold body, radiation of heat, but some simple processes like production of heat by friction cannot be explained by it. This theory has now been replaced by theory of molecular motion.

Heat Capacity. The amount of heat needed to raise the temperature of a body through one degree Celsius (symbol: C). It is measured in joules per kelvin. For a gas, the heat capacity may be measured under conditions of either constant pressure or constant volume.

Heat Death. The condition of any isolated system when its entropy is a maximum. If the universe is a closed system it should eventually reach this state. This is called the *heat death of the universe*.

Heat Engine. A device that converts heat energy into work. Examples are petrol and Diesel engines, gas turbines, and steam engines. In terms of *thermodynamics,* the heat engine converts a quantity of heat q_1 at temperature T_1 into work *w*, but the conversion can never be complete, some of the heat q_2 will be wasted and can be regarded as being discharged into a heat sink at a temeperature T_2. The efficiency of the engine is defined as the work output divided by the heat input (w/q_1). As the work, *w*, is equal to q_1—q_2 (according to the first law of thermodynamics), the efficiency is $1\text{-}q_2/q_1$, or $1\text{-}T_2/T_1$. It is therefore important to make T_1 as high as possible and T_2 as low as possible.

Heat Latent. The amount of heat required to convert unit mass of a susbstance from one physical state to another, e.g., at constant temperature; to convert 1 gm. of ice into water or vice versa, or to vonvert 1 gm of water into steam, etc.

Heat Pump. A device that extracts heat from one substance at a low temprature and supplies it to another susbtance at a higher temperature. Such a process would violate the second law of *thermodynamics* if it occurred spontaneously; therefore the heat pump necesarily consumes energy in the process. Heat pumps are used to extract the low temperature heat from rivers that flow through towns and to convert it to a higher temperature so that use can be made of it, for example in space-and water-heating units.

Heat Transfer. The transference of energy between two bodies or regions by virtue of the difference in temeperatures between them. The three methods of transference are: *convection, conduction* and *radiation.* In convection the heat is transferred by a hotter region flowing into a colder region, either as a result of density differences (natural convection) or by using a fan (forced convection). In conduction, on the other hand, the heat is transferred by direct contact without any apparent relative motion. In radiation the heat is transferred by means of either *infrared* radiation or *microwave* radiation. Radiation is the ony method of transferring heat through a vacuum. The *Dewar* (Thermos) flask is designed to minimize heat transfer whereas the heat exchanger or radiator is designed to maximize it.

Heaviside Layer. Heaviside suggested that an ionized layher in the air, concentrate with the earth's surface, might serve as a reflecting surface which would confine radiation between it and the earth. The existence of such a layer, or rather system of layers, has been conclusively proved. The two main levels have been given the names 'Heaviside' or 'Kennelly-Heaviside layer' for the lower layer and the '*Appleton Layer*' for the upper layer. More recent practice has been to call them the E and F layers respectively.

Heaviside-Lorentz System of Units. C.g.s. (centimetre, gram, second) units in which the force between two magnetic poles m_1 and m_2 a distance d apart in medium of permeability is m_1m_2/d_2 and in which the force between two charges q_1 and q_2 a distance r apart in a medium of permittivity ε is $q_1q_2/\varepsilon r_2$.

Heisenberg Force. An exchange force between necleons in which charge (equivalent to spin plus position) is excharged.

Heisenberg's Uncertainty Principle. It is impossible to measure simultaneously and accurately the position and momentum of a subatomic particle. If such a measurement is carried out simultaneously, the uncertainty in the measurement of position (Δx) and momentum (Δp) is given by the equation

$$\Delta x \ \Delta p = \frac{h}{4\pi} \quad \text{where } h \text{ is Planck's constant.}$$

Helmert's Formula. An empirical formula which provides a value for the gravitational acceleration, gΦ, H at any latitude Φ and altitude H (in metres) $g\Phi, H = 9.80616 - 0.025928 \cos 2\Phi + 6.9 + 10^{-5} \cos^2 \Phi - 3.016 \times 10^{-6} H$.

Helmholtz. A unit of dipole moment per unit area equal to 3.335×10^{-10} Cm^{-1}

Helmholtz Coil. Refers to a pair of similar coils designed to produce unifrom field in the region between them when a current is passed through them. They are connected in series and are placed parallel to each other at a distance equal to their radius.

Helmholtz Function (F). F = U - TS, where U is the internal energy, T the absolute temperature and S the entropy of the system F is a state function.

Helmholtz's Relation (Optics). If α_1 and α_2 are the inclinations of light rays to an axis passing through a spherical surface, where the refractive indices on the respective sides of the surface are n_1 and n_2 then the object and image heights y_1 and y_2 are related by

$$y_1 n_1 \tan \alpha_1 = y_2 n_2 \tan \alpha_2.$$

The equation is only valid for small angles and may be approximated by

$$y_1 n_1 \tan \alpha_1 = y_2 n_2 \alpha_2.$$

Usually in this form it is referred to as *Lagrange's Law.*

Helmholtz Reasonators. A set of vessels, each resonating at particular frequency. They can be used to detect the frequency at which each reasonates.

Helmholtz's Theorem (for Fluids). In the absence of body forces (when entropy is a constant), individual vortices in a non-viscous fluid always consist of the same fluid particles.

Henry (H). The **SI* unit of inductance equal to the inductance of a closed circuit such that a rate of change of current of one ampere per second produces an induced e.m.f. of one volt. Named after Joseph *Henry.*

Hertz (Hz). The *SI* unit of frequency equal to one cycle per second. Named after Heinrich *Hertz.*

Hertz Effect. The enhancement of a spark discharge by passing ultraviolet light to promote ionization.

Hertzian Waves. Electromagnetic waves of frequencies between zero and approximately 10^4 MHz.

Heterogeneous Radiation. A particular type of radiation like X-rays or gamma rays, having a variety of wave-lengths or quantum energies.

Heterogeneous Reactor. A type of nuclear reactor wherein the fuel is separated from the moderator.

Heteropolar Generator. A generator wherein the active conductors pass through magnetic fields of opposite sense in succession. An alternating e.m.f. is induced in each conductor so that a direct current generator of this type must be fitted with a commutator.

Hexode. A thermionic valve having four grids between the cathode and anode, generally empolyed as a frequency-changer.

High Frequency. A radio frequency in the range between 3×10^6 to 3×10^7 Hz.

Hiltner-Hall Effect. The polarization of light from distant stars. It is thought to occur in interstellar regions and to be due to the selective absorption of light vibrating in one plane by ferromagnetic grains which are aligned in a magnetic field in these regions.

Hole. A vacancy due to a missing electron in a crystalline semiconductor. When an electron from a neighbouring atoms fills this vacancy, a new vacancy or hole is created at the neighbouring atom. This gives rise to hole conduction. A hole can be regarded as having a positive charge and mass.

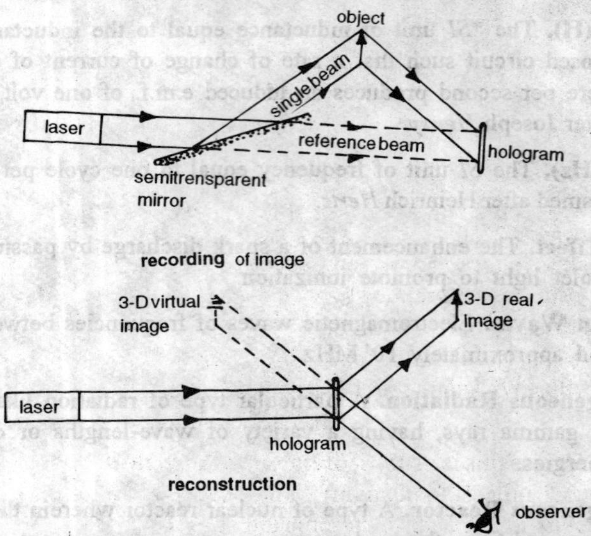

Holography. **A three-dimensional image formed by two beams of light is recorded as an interference pattern on a single plate. The two images, giving a 3-D effect, are reconstructed by shining a similar beam through the hologram.**

Holography. A method of producing a stereoscopic image without using a camera. A monochromatic beam of *coherent* radiation from a laser is split into two using a semitransparent mirror; one beam falls directly onto a photographic film or plate and the other is reflected by the subject onto the film. The two beams form *interference* patterns on the film, which is called a hologram. To reconstruct the image, light of the same wavelength from a laser is shone onto the hologram. The interference pattern on the hologram diffrancts the beam and splits it into two parts. One part gives a real two-dimensional image and the other a virtual three dimensional image. The theory of holography was suggested by Dennis *Gabor* in 1947 but could not be put into practice until the laser was invented 16 years later.

Hooke's Law. For an elastic body, the *stress* is directly proportional to the *strain*. For example, if a heavy mass is hung from a wire, the fractional extension is directly proportional to the mass. The law applies only up to the elastic limit (*see* elesticity). Named after its discoverer Sir robert *Hooke,* who stated it in the form *ut tensio, sic vis.*

Horizontal Component of Earth's Magnetic Field (H). The value of the Earth's magnetic field in a horizontal direction at a given point on the Earth's surface. Its value varies from place to place.

Horse-Power. (H.P.). A unit of power in the f.p.s. system equal to 33000ft. lb per min. or 746 watts.

Hot. Highly radioactive. A *hot atom* is an atom in excited state or having kinetic energy above the thermal level of the surroundings, generally as a result of nuclear processes.

Hot-cathode Tube. A discharge tube where in the supply of electron for carrying the discharge is provided by thermionic emission from a heated element.

Hot-wire Gauge. A pressure guage depending on the cooling by the gas of a hot filament.

Hot-wire Microphone. A device for the measurement of the intensity and amplitude of sound waves, based on the decrease in the resistances of an electrically heated wire on being subjected to sound waves.

Hubbard Model. This is a highly simplified model which attempts to account for magnetic ordering in metals. In the model the large number of bound and continuum electron levels associated with each ion is reduced to a localized orbital level. The level of each ion is specified as empty, or containing one electron of one or other spin, or two electrons of opposite spins, and the Hubbard Hamiltonian derived on this basis. This is found to contain two types of tems. The first type gives rise to local magnetic moments; the second type gives rise to a band spectrum. When both types of terms are present, exact analysis is not possible, but it can predict in certain cases a transition from a non-magnetic metal to an antiferromagnetic insulator. The latter is sometimes referred to as a Mott-Hubbard Insulator.

Hubble's Law. The ratio of velocity to distance of galaxies in the universe is a constant. The constant is called *Hubble's constant.* Its reciprocal is the time calculated to have elapsed since the collapsed state of the universe existed and is called *Hubble time.* This time is thought to be approximately 1.9×10^{10} years. The Hubble constant may increase at extreme distances because of a decrease of recessional velocities under the influence of gravity.

Hum. Extraneous alternating currents that appear in the output of the amplifier having their origin in an associated or nearby electric power circuit.

Humidity. A measure of the amount of water vapour in the atmosphere. Absolute humidity is the mass of water vapour in unit volume of air, measured in kilograms per cubic metre. Relative humidity is the ratio of the absolute humidity at a given temperature to the maximum humidity without precipitation at the same temperature, usually expressed as a percentage.

Hugen's Construction. Every point of a wave front can be regarded as a source of secondary wave lets knowing the position of the wave front at any item. t_0, its position at any subsequent time t is obtained by constructing secondary wave lets by taking different points on the wave front as the centre and drawing spheres with radii equal to $v(t - t_0)$ where v is speed of the wave in the medium. A surface tangential to all these secondary wave lets then gives the position of the new wave front.

Hygen's Principle (Huygens-Fresnel Principle). Every point of a wave can be considered as the centre of a new elementary wave. The resultant wave produced by the interference of all these elementary waves is identical with the original wave.

Hunting. Variation of temperature of a thermostat above and below the desired value. It can be eliminated by damping.

Hydraulic Press. A machine which is based on Pascal's law. In this device, a force applied over a small area by a piston is transmitted through fluid to another piston having a large area. Thus a small force may be used to obtain a large force.

Hydraulics. The study of the applications of *hydrostatics* and *hydrodynamics* to design problems. In civil engineering it is used to study the flow of water in pipes, rivers, canals, etc., especially with reference to the construction of dams, reservoirs, and hydroelectric power stations. In mechanical engineering, applications include the design of machinery involving fluids, such as hydraulic presses, *turbines,* propellers, etc. Hydraulics is concerned with the bulk properties of fluides, such as density, viscosity, elasticity, and surface tension, rather than their molecular properties.

Hydraulis (Greek: Water Pipe). An early type of *organ* in which the air pressure was maintained by water. A clay model found in the ruins of Carthage has three ranks of pipes and suggestes an actual height of about 3 m (10 ft). The loud sound it produced made it useful for signalling in battle; it was also played in Roman amphitheatres, Emperor Nero being a keen performer.

Hydroelectric Power. Electricity generation using the energy of falling water. The water turns a *turbine* connected to an alternator, generating electricity with an efficiency of over 90% at full load and generally over 60% at quarter-load. Water is led through pipes from high-level natural or artificial reservoirs to the power station. Lower-level reservoirs and dammed rivers are also used in some situations. The higher the reservoir, the less water is needed for the same power output. Hydroelectric power is, therefore, a cheap power source in mountainous areas with high rainfall. Unfortunately these are not usually near the industrial communities that consume the most power. Also, because it depends on rainfall, hydroelectricity has to be backed by other power sources. In pumped storgage stations, electricity is stored by using it to drive pumps that raise the water to a high-level reservoir. In times of high demand this water is run back through the turbines. Hydroelectric power stations can have an output of 10 000 megawatts but in the UK most produce less than 100 MW.

Hydrometer. An instrument for measuring the relative *density* of a liquid. It consists of a weighted and sealed glass bulb with a long neck on which a scale is calibrated. The relative density is measured by placing the hydrometer in the liquid and noting the level to which it sinks.

Hydrophone. A type of *microphone* that converts under-water sound waves into electrical signals. It consists essentially of a diaphragm, a *transducer,* and an *amplifier.* It is used in *sonar* equipment, depth sounding, and underwater communications.

Hygristor. An electronic component with an electrical resistance that varies with humidity.

Hygrograph. An instrument used for measuring and recording the relative humidity of the air.

Hydrometer. An instrument that measures the relative *humidity* of the atmosphere. In mechanical hygrometers, a material (usually human hair) is used, the length of which varies with the humidy; the variations being transformed into the movement of a pointer along a scale. In the wt-and-dry bulb hygrometer, two thermometers are placed side by side, one having its bulb covered by a moist cloth. The coooling caused by the evapotation from this wet bulb depends on the atmospheric moisture and thus the difference between the two thermometer readings can be related by standard tables to the relative humidity.

Hygroscope. An instrument that indicates variations in the relative humidity of air. It uses a substance that changes colour in the presence of *moisture.*

Hypermetropia (long sightedness). Defect of the eye to see near objects distinctly. It can be corrected by using convex lens spectacles.

Hypsometer. An apparatus used for measuring the boiling point of a liquid. Since the boiling point of a substance depends on the altitude, it can be used for the measurement of altitude above sea level. Also used for caliberating thermometers at steam temperature.

Hysterisis (Magnetism). If a ferromagnetic substances is magnetized by an external applied field H and then demagnetized, the flux density lags behind the field H. It also takes place in the elastic behaviour of materials. If a body is subjected to a stress, it is strained. When the stress is completely removed, the strain does not disappear completely at once. This lagging behind of strain is known as hysterisis.

Hysterisis Cycle (Hysterisis Loop). A closed loop which is obtained when an initially demagnetized ferromagnetic substance is subjected to a cycle of magnetic field variations. The unmagnetized ferromagnetic substance is first subjected to a magnetic field till it reaches magnetic saturation (point A). The field is then changed as shown in the figure and the corresponding magnetic flux observed. The area of the loop formed during one cycle is equal to the energy loss in taking the sample once through the cycle. This is known as hysterisis loss. The magnetic flux density remaining is the sample when the saturating field is removed is known as the remanence (corresponding to the point C). The value of magnetizing field strength required to bring this remaining flux density to zero is known as the *coercivity* of the substance. This corresponds to the value O D in the curve.

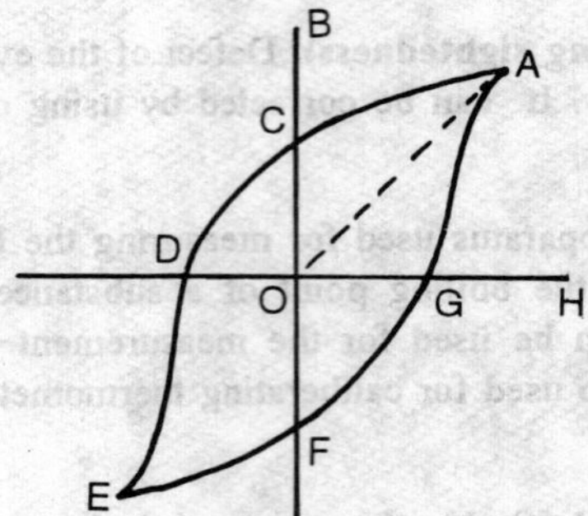

Hyteresis Loss. (1) (magnetic) The dissipation of energy that occurs on account of magnetic hysteresis, when the magnetic material is subjected to changes of magnetization. (2) (dielectric) The dissipation of energy that occurs, due to dielectric hysteresis when the dielectric is subjected to a changing electric force (3) (elastic). The dissipation of energy through torsional hysteresis.

I

Illuminance of a Surface. Refer to the amount of luminous energy falling on unit area of the surface per unit time. Its unit are lumen / m^2 (or Lux).

Image. A point or region where rays of light from an object meet or appear to meet after reflection or refraction from a reflecting or refracting surface. The image is said to be real. If the rays of light actually meet and is called virtual if they appear to diverge from a point.

Immersion Objective. An objective used in high-powered microscopes. The space between the lens and the microscope slide to be examined is filled with suitable liquid. Ceelar-wood oil is the liquid generally used. The arrangement increases the effective aperture of the objective and causes more light to enter the system than in the absence of oil.

Impedance (Z). It is a measure of the resistance offered by a circuit to alternating curent. It is equal to sum of the resistance R and the ractance X of the circuit where X is

$$L^2 \omega^2 + \frac{1}{c^2 \omega^2}$$

L, C, ω denote the inductance, capacitance and angular frequency respectively.

Impulse. The product of the force and the time during which the force acis. It is equal to the change in the momentum of the body.

Incandescence. The emission of light by a surface on being heated to a high temperature.

Incandescent Lamp. An electric lamp wherein light is produced by the heating effect of a filament of carbon, osmium, tantalum, or tungsten.

Incidence (Angle of). Refers to the angle between the ray striking reflecting or refracting surface (i.e. the incident ray) and the normal to the surface at the point of incidence.

Inclination. (Angle of dip) (δ) the angle between the Earth's magnetic field and the horizontal at any point on the Earth's surface.

Inclinometer. An instrument used for measuring the magnetic inclination.

Incoherent. Not coherent. Radiation from any source other than a laser is incoherent.

Index Error. *Syn.* zero error. A scale error on a measuring instrument so that the instrument shows a reading x when it should show zero reading.

Indicator Diagram. It is the cycle traced out during the motion of a piston in the cylinder of an engine. Vertical displacements represent the pressure and horizontal displacements the volume of the working substance. The area enclosed by the diagram gives the work done per cycle and is used in estimating the efficiency of the engine.

Indicator Tube. A minute cathode-ray tube in which the shape or size of the image on the screen changes with the input voltage V in such a manner that it is used to measure V, and therefore can indicate the value of a changing signal.

Induced Current (Induced e.m.f.). See Faraday's Laws of e.m.f. Induction.

Inductance (L). E.M.F. induced in a circuit on account of a changing electric current in the circuit or in a neighbouring circuit when the rate of change of current is unity. It is measured in henry.

Induction, Electromagnetic. Development of an e.m.f. induced in a circuit when the magnetic flux linked with the circuit varies.

Induction Coils. An instrument for production of high voltage impulses from a source of low e.m.f. It is made up of a soft-iron core over which two coils—a primary and a secondary are wound. The number of turns in the secondary coil is much larger than in the primary. A large pulsed e.m.f. can be produced in the secondary by a pulsed curent in the primary that can be produced by a rapid variation of the electric current in this coil.

Induction Motor. An electric motor in which an A.C. that is passed through the primary winding induces electric current in the secondary of the motor. The magnetic interaction between the two coils causes the rotor to turn.

Inertia. The property of a body by virtue of which it opposes any change in state of rest or of uniform motion. The property is implied by Newton's first law of motion.

Inertial Mass. The mass of a body as determined by its inertial property. When a body is accelerated by a constant force F and the accelerated produced in it is 'a', then the inertial mass of the body

$$m = \frac{F}{a}$$

Inertial System (Inertial Frame of Reference). Frame of reference for which acceleration is zero. Two frames of reference are inertial with respect to each other if there is no relative acceleration between them.

Infrared Rays. Radiation of wave length lying in the range 7×10^{-5} cm to 10^{-1} cm.

Irfra Sound. Sound of pressure waves having frequency below 20 Hz.

Infrared Telescope. A reflecting *telescope* for detecting and studying infrared radiation from astronomical sources, such as stars enveloped in dust clouds or cool *molecular* clouds of gas and dust. Semiconductor detectors are used, and these (and in some cases the telescope optics) must be cooled to very low temeperatures. Infrared telescopes are normally sited at high altitudes to reduce infrared absorption by the atmosphere or are carried on satellites. Several major telescopes are located at the summit of Mauna Kea (4200 m) in Hawaii.

Infrasonics. The study of sound waves with a freequency below the threshold of human hearing, i.e. below about 20 hertz. they are produced naturally by earthquakes and tidal waves; in cars tavelling at high speeds they have been suspected of having adverse effects on the drivers.

Instantaneous Value. The value at any paticular instant of time.

Insulator. A substance that does not conduct electricity e.g. wood plastic etc.

Integrated Circuit (IC). A solid-state *semiconductor* circuit contained in a single wafer of semiconductor. ICs are made by a process of etching and diffusing a pattern of impurities into the semiconductor surface, forming tiny p-n junctions, which make up individual diodes, *transistors,* etc. Since the 1970s computers have been used to make ICs smaller and more complex, in spite of the fact that further reducing the size makes the doping pattern imprecise. A silicon Chip (or microchip) is covered with circuits, and computer-controlled microscopic probes search out the best points for connections for each specific device. The application of this technique has revolutionized many areas of industry and commerce, making cheap microprocessors available for the automation of production processes.

Integrating Galvanometer. A galvanometer wherein inertia and control of the moving system are so chosen that the change in flux produced in an exploring coil can be measured even through the changes last over a period of several minutes.

Integrator. A mechanical or electrical instrument for performing the mathematical operation of integration, e.g. the a.c. current *i*, flowing into a capacitor C gradually builds up a voltage on the capacitor equal to $1/C \int i\, dt$, i.e., an integration of *i* with respect to time is performed.

Intensifying Screen. A screen, coated with calcium tungastate crystals, it emits light under the action of X-rays and reduces the exposures needed in radiography by a factor of about 100.

Interference. A wave phenomenon in which two waves combine either to reinforce each other or to cancel each other out, depending on their relative phases. The pattern of light and dark strips so produced is called an *interference pattern* (or interference fringe).

The effect occurs when the two beams have the same frequency and have approximately the same amplitude. Interference was discovered in 1801 by Thomas *Young and provided strong evidence for the wave theory of light.

An *interferometer* is used to produce interference patterns, mainly for the accurate measurement of wavelengths. Several different types are in use, the older instruments being devices for splitting a beam into two parts and then recombining them to form interference patterns after each part has travelled a diffferent distance. In the Michelson interferometer (Michelson-Morley experimeet) the beam is split by a half-silvered mirror in which part of the beam is reflected and the rest transmitted. In the Fabry-Perot interferometer two parallel half-silvered mirrors are placed close to each other. In more modern instruments two lasers are used as separate coherent sources.

Intermolecular Forces. Forces between molecules and atoms. These force are attractive in nature.

Internal-Combustion Engine. A **heat* engine in which fuel is burnt inside the engine, rather than in a separate furnace. This category includes all *piston* engines, *jet engines* and *rockets.* The first practical internal-combustion engine was patented by N. *Otto* in 1876. This was a four-stroke engine using gas as a fuel. It was the invention of the carburettor and the development of the *oil industry* that made the liquid-fuelled Otto engine a source of power for the horseless carriages emerging in the late 19th century.

The modern petrol engine has a compression ratio of 8 or 9 to 1, which requires special fuels to avoid *knocking* although lead pollution is causing manufacturers to design engines suitable for lead-free petrol. A simpler but less efficient variety of the petrol engine is the two stroke. This does not have the complicated inlet and exhaust valves of the Otto engine, the explosive mixture entering and leaving the cylinder through ports in its walls that are covered and uncovered by the movements of the piston. Two-stroke engines are used where low power is required and in some motorcycles.

The main alternative to the petrol engine is the oil engine, based on a cycle invented by the German engineer Rudolf Diesel (1858-1913). In the Diesel engine, air is compressed alone inside the engine, causing its temperature to rise to over 550°C; oil is then pumped into the combustion chamber as a fine spray and ignites on contact with the hot air. In this case the compression ratio has to be 15 or 16 to 1, making the engine considerably heavier and more expensive than the petrol engine. The efficiency of both Otto and Diesel engines is limited by their compression ratios and thus their working temperatures (heat engines). Also, combustion is intermittent and therefore incomplete, causing pollution problems. Moreover, both being reciprocating engines, they have inherent vibrations. This last problem is overcome to some extent in the *Wankel engine*. The *gas turbine,* however, uses continuous combustion and with a compression ratio of up to 30: 1 can reach of working temperature of 1200° C. It is therefore more efficient and creates less pollution than piston engines. Jet engines based on the gas turbine are used in aircraft and to provide an easily started prime mover in some power stations. Because all forms of internal-combustion engines rely on fossil fuels they are prodigious producers of carbon dioxide.

Internal Energy (U). The total energy of a system, that is a function of its temperature T, and is the sum of the kinetic and potential energy of its molecule. It increases when the temperature of the system is increased or work is done on the system externally.

Internal Resistance (r) of a Cell. Resistance offered to the flow of current in a cell. Due to the internal resistance of the cell, its potential difference V is less than its e.m.f. (E). The two are related as

$$E = V + ir \text{ or } V = E - ir$$

where i is the current drawn from the cell.

Internal Work. The work done in separating the molecules of a system against their force of attraction. Its value is zero for an ideal gas.

International Atomic Energy Agency. A specialized agency of the UN established in 1957 to promote peaceful applications of atomic energy. Based in Vienna, it stages an influential annual conference.

International Atomic Time. The scientific standard of time derived from the *SI* unit, the second, which is based on the *caesium clock*. By means of atomic clocks and satellites, the major nations of the world are able to synchronize their time standards.

International Candle. A unit of luminous intensity.

International Steam Table Calorie (I.T. Calorie). A heat unit fomerly adopted by international agreement as being 4.1868 x 10^7 ergs to avoid errors due to the variation of the specific heat capacity of water with temperature. It is now replaced by the joule; 1 cal_{it} = 4.1868 joules.

Interpolation. Estimation of the value of a function, *y*(*x*), for a value of the variable, *x*, that lies between those for which the function is known. This may be done graphically or by using an interpolation formula.

Intrinsic (i-type) Semiconductor. A pure semiconductor wherein the electron and hole densities are equal under conditions of thermal equilibrium.

Inverse Gain of a Transistor. The gain of a bipolar junction transistor when it is connected with the emitter as the collector, and vice versa.

Inverse Square Law (Light). The illuminance of a surface is inversely proportional to the square of its distance from the (point) source The law, holds for isotropic sources and at distances large compared to the size of the source.

Inverter. (1) Any device that charges d.c. into a.c. (2) *Linear inverter*. An amplifier that inverts the polarity of a signal i.e. introduces a 180° phased shift. *Digital inverter*. A logic circuit whose output is low when the input is high and vice versa.

Loffe Bars. Heavy current carrying bars used in experimental-fusion devices to increase plasma stability.

Ionization. The process of formation of ions by ionizing radiation on collision with fast moving particle or by chemical reactions.

Ionization Chamber. An instrument used for measuring the intensity of *ionizing* radiation. It consists of a gas-filled chamber containing two electrodes with a large potential difference between them. When radiation enters the chamber it ionizes some of the gas atoms or molecules. The ions flow towards the electrodes creating a current, whose magnitude is a measure of the intensity of the radiation.

Ionizing Radiation. Any radiation that ionizes the atoms or molecules of matter. It may consist of particles (such as *electrons*) or it may be electomagnetic. Ionizing radiation occurs naturally and is emitted by radioactive substances. It is produced artificially in X-ray machines, particle accelerators, nuclear reactors, etc. Ionizing radiations are used in medical diagnosis and therapy and in sterilization of food. In biological tissue these radiations create reactive free radicals, especially by ionizing water molecules, which attack proteins, nucleic acids, etc., and can cause damage by changing their structure and function. This is how they kill cancerous cells, which are more susceptible than normal cells. Higher doses can change normal cells into cancerous cells and cause other illnesses. However, man has evolved in an environment in which there is a backgroud of radiation and is therefore adapted to a certain level of it. On average we each receive about 2150 microsieverts from the air (37%), the earth (19%), within the body (17%), *cosmic* rays (14%), medical X-rays (11.5%), fall-out from nuclear tests and air travel (0.5% each), industry (0.4%), and *radioactive* wastes (0.1%).

Ionosphere. A region of the upper atmosphere that reflects short radio waves, enabling transmissions to be made round the curved surface of the earth by sky waves. The gases in the ionosphere are ionized by absorption of radiation from the sun. Its existence was suggested in 1902 by A.E. Kennelly and independently by O. Heaviside. Sir Edward Appleton (1892-1925) provided proof by bouncing radio waves off the different layers of the ionosphere, which vary in behaviour with the position of the sun and with the sunspot cycle. The ionization of the D region, at between 50 and 90 kilometres altitude, disappears during the day. The E region is between 90 and 160 kilometres high and the F region (sometimes called the Appleton layer) is from 160 kilometres up to about 400 kilometres. The lower part of the ionosphere (E region) is sometimes called the Heaviside-Kennelly layer. The gas particles in the F region do not lose their charge as quickly as those below because the gas is less dense and therefore ions and electrons are subject to fewer collisions. This enables radio transmissions to continue at night.

Irreversible Process. A process which cannot be returned to its original state through the same series of stages. In such processes, the system is not in equilbrium throughout the change. All real process are irreversible in nature.

Isenthalpic Process. A process taking place without any change of enthalpy i.e. so that the totla heat energy remains constant.

Ising Model. Used, for instance, if feromagnetism and phase transformation theory, it is a model for a system of interacting particles. A scalar variable, σ associated with each lattice point of a crystal has possible values + 1 or -1 representing the two directions of electron spin in a ferromagnet or the two types of atom in a binary alloy. The interaction energy is given by

$$E_{ij} \begin{cases} = -j\sigma_i\sigma_j \text{ where } i, j \text{ are nearest-neighbour particles} \\ = 0 \text{ where } i, j \text{ are not nearest-neighbours.} \end{cases}$$

J is a coupling constant which can be positive as in a feromagnetic system, or negative as in an antiferromagnetic system. The model enables a value for the heat capcaity to be calculated; the two-dimensional form and the three-dimensional form (which has not been solved exactly) lead to a transition point.

J

J. The symbol represents the mechanical equivalent of heat.

Jansky Noise. Refers to high-frequency static disturbance of cosmic origon.

Jet Engine A form of *gas* turbine in which part of the energy released by burning the fuel drives a turbine, which turn drives a compressor to in crease the pressure of the air required for combustion, and part is used as a high-velocity jet to provide thrust to drive an air-craft. The jet engine was patented in 1930 in the UK by Sir Frank *Whittle,* the first practical Whittle engine powering a Gloster aircraft in 1941. Except for light aircraft, most military and commercial aircraft are now powered by jet engines. Thrust in these engines is equal to the mass of the gas produced multiplied by its acceleration and is due to the pressure of the expanding gas on the engine itself rather than on the air through which it is flying. It is therefore more efficient at higher altitudes, where the atmosphere is thinner.

Early postwar commercial aircraft (e.g. the Vickers Viscount) used a *turboprop* engine, in which a propeller is driven by the turbine shaft. For greater speed and fuel economy the turboprop has been replaced by the *turbojet,* in which the turbine drives one or more compressors but most of the energy of combustion is used to provide *jet propulsion.* For bursts of extra energy, especially in military aircraft, extra fuel can be burnt in the exhaust gases (called *reheat* or *afterburing*). In the *turbofan* a large fan, driven by an inner shaft from the turbine, precedes the usual compressor and compresses the cool air, which by passes the combustion chambers and mixes directly with the exhaust jet, giving greater fuel economy and quietness. These are used on modern subsonic jumbo jets.

At over twice the speed of sound (Mach 2), the forward pressure of the air is sufficient for the compressor, and therefore the turbine, to be dispensed with. The resulting engine is called a *ramjet* or, colloquially, a flying drainpipe. The main drawback of the ramjet is that it needs a rocket-assisted take-off.

Turboshaft engines, similar to the turboprop, are also in use for driving helicopters, hovercraft, ships, trains, turbogenerators in power stations, and (experimentally) cars. In these the turbine shaft is used to provide all the working power.

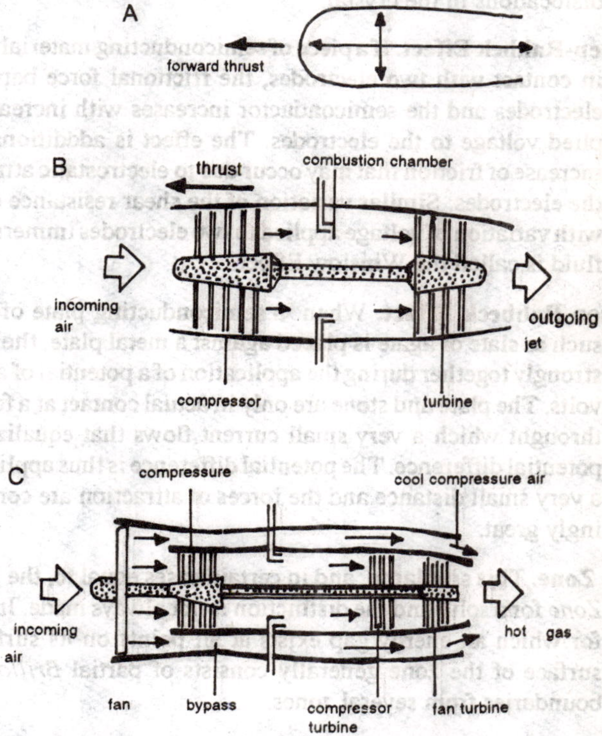

Jet Engine. (a) The principle: equal pressure from the expanding gas inside the-engine meets equal resistance except rearwards. The result is a forward thrust. The lower the outside pressure, the greater the thrust. (b) The turbojet, used for subsonic and supersonic flight (especially with reheat). (c) The turbofan, used for subsonic flight: cool air, compressed by the fan, by passes the engine and mixes with the hot jet.

Jet Propulsion. Propulsion where in the propulsive force is the reaction of one or more jets of hot gases coming out at high velocity from backwardly directed nozzles.

Jet Tones. The unsteady tones formed when a stream of air is projected into stationary air from an orifice. The moving fluid tends to curl outwards into the stationary fluid that forms alternate vortices on each side of the jet.

Joffe Effect. An observed decrease in brittleness or rock salt on dissolution of the surface. The effect is probably associated with the dislocations in the crystal.

Johnsen-Rahbek Effect. If a piece of semiconducting material is placed in contact with two electrodes, the frictional force between the electrodes and the semiconductor increases with increase of applied voltage to the electrodes. The effect is addditional to any increase of friction that may occur due to electrostatic attraction of the electrodes. Similar variation of the shear resistance of a fluid with variation of voltage applied to two electrodes immersed in the fluid is called the Winslow Effect.

Johnson-Rahbeck Effect. When a semiconducting plate of material such as slate or agate is placed against a metal plate, the two hold strongly together during the application of a potential of about 200 volts. The plate and stone are only in actual contact at a few points throught which a very small current flows that equalize a high potential difference. The potential difference is thus applied across a very small distance and the forces of attraction are correspondingly great.

Jones Zone. This similar to, and in certain cases equal to, the *Brillouin Zone* for a solid and the distinction is not always made. It is a zone for which an energy gap exists at all points on its surface. The surface of the zone generally consists of partial *Brillouin zone* boundaries from several zones.

Josephen Effect. Effect observed in two superconductors which are separated by a thin dielectric when a steady potential difference V is applied. An oscillatory current is set up having a frequency proportional to V.

Joshi Effect. A fall (negative effect) or rise (positive effect) in the low-frequency alternating current passing through a gas dielectric condenser when the gas is irradiated continuously with visible light.

Joule (J). Unit of work in the SI system. 1 Joule is equal to the work done by a force of 1 Newton when it moves an object through a distance of 1 meter.

Joule Effect. A rod of para-or diamagnetic material suffers a mechanical change in length when subjected to a magnetic field.

Joule's Law. The heat produced by an electric current I flowing through a resistance R for fixed time t is given by $I^2 Rt$. If the current is measured in amperes, the resistance in *ohms*, and the time in seconds, then the heat produced is in joules.

Joule-Thomson Effect (Joule-Kelvin Effect). When a gas is allowed to expand adiabatically through a porous plug, the temperature of the gas changes. The rate of change of temeperature with pressure on adiabatic expansion, known as the *Joule-Thomson coefficient*, may be taken to be constant if the pressure difference across the plug is small. The differential Joule-Thomson effect, $\Delta T/\Delta p$, where ΔT is the change in temeprature for a small change in pressure Δp at temperature T, can be shown from thermodynamical considerations to be given by the expression

$$\Delta T = \frac{T\,[\,\delta V / \delta T\,]_P - V}{C_P} \quad \Delta p$$

where V is the volue of the gas and C_p is the specific heat at constant pressure. The sign of the Joule Thomson coefficient will depend upon the relative values of the terms in the numerator of the above expression. The value of the numerator is dependent upon the temperature, and there is a fixed temperature at which the sign of the Joule-Thomson coefficient changes. This temperature is known as the Joule-Thomson inversion temperature. Most gases are below their inversion temperatures at room temperature and hence may be liquefied by expansion. The so-called permanent gases (e.g. hydrogen. $T_1 = 190$ K), have an inversion temperature below room temperature and must be cooled before they can be liquefied by this means.

Joule's Equivalent of Heat (J). A constant that relates unit of heat to the unit of mechanical energy.

1 calorie = 4.2 joule.

The value of J = 4.2 joule/calorie.

Jurin Rule. The height of rise of a liquid in a capillary tube of radius α is given by :

$$h = \frac{2\gamma \cos\alpha}{\rho g a}$$

where γ is the surface tension of the liquid, α is the angle of contact with the capillary and ρ is the density of the liquid.

Juvin's Rule. When a capillary tube of radius *r* dips vertically in a liquid having density *p* and surface tension *y* the liquid rises through a distance *h* up the tube, given by

$$h = 2\gamma \cos \alpha / rgp$$

where α is the angle of contact between the liquid and the walls of the tube and *g* is the acceleration of the free fall. For a liquid that does not wet glass α exceeds 90^0 and *h* is negative: the liquid is depressed in the bore.

K

Kaiser Effect. Very low-energy acousitic pulses, which are observed when a metal is initially deformed, and which are not observed on reapplication of any stress up to the original limit. The effect arises form workhardening removing the plastic region of the stress-strain curve. Acousitic emission reappears upon reversal of the stress or upon application of a stress exceeding the original limit.

Kane Band. A form of energy band which commonly occurs in semiconductors which have a narrow gap between the valence and conduction band. For such bands, if the energy of the electrons (or holes) is experessed as a function of wave-vector, the bands are very 'skinny'. The effective mass is a function of the electron or hole energy in the band and is very small for small electron (or hole) energies.

Keeper. Syn. armature. The regions near the ends of a magnet produce an induced flux within it that opposes the original magnetizing flux, the effect being maximum for the shortest magnets. The flux included in the keepers is opposite in direction to that in the magnet itselft, and so the demagnetizing effect in the magnet is neutralized.

Kellner Eyepiece. A type of eyepiece having a cemented eye lens that corrects chromatic aberration and distortion. It is commonly used as an eyepice in prism binoculars.

Kelvin. Symboll: K. The SI unit of termodyamic temperature. It is 1/273.16 of the thermodynamic temperature of the triple point of water.

Kelvin Balance. A type of abalance that consists of six coils, four fixed and two which move between them on a balanced rod. The suspension, is made up of two flexible multiple copper ribbons that serve to carry current to the coils, so that each fixed coil tries to displace the balanced arm in the same direction when current flows. The instrument can measure d.c. and a.c. .and cna also read wattage.

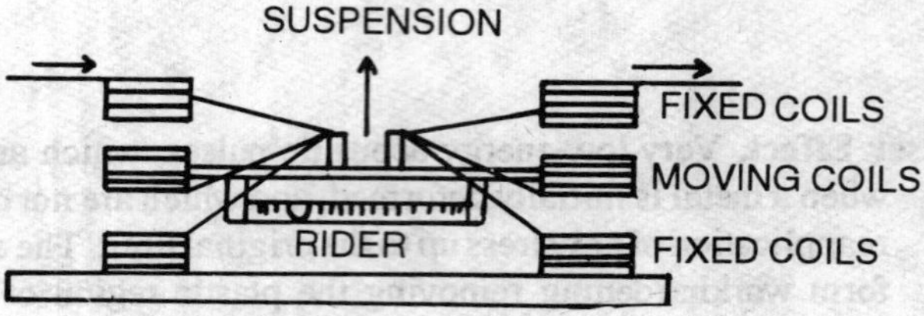

Kelvin Contacts. A method used for testing or measuring electonic circuits or components. Double leads are used to test points, one lead carrying the test signal and other lead going to the measuring instrument. It removes the effect of the resistance of the leads on the measurement.

Kelvin Double Bridge. A special development of the d.c. Wheatstone bridge for measurement of low resistances. A is the low resistance to be measured and B is a known low resistance of the same order. R_1/R_2 is kept equal to r_1/r_2 but they are changed until no current can be detected by the galvanometer, G. Then,

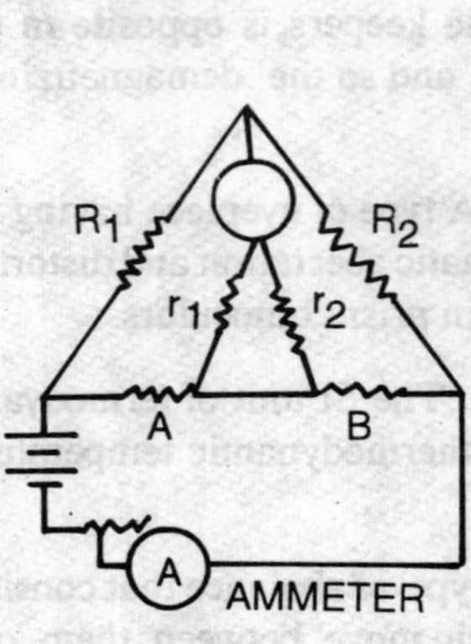

$$\frac{A}{B} = \frac{R_1}{R_2} = \frac{r_1}{r_2}$$

Kelvin, Degree (k) SI Unit of Temperature. The Kelvin scale of temperature is based on the triple point of water, which is assigned the temeperature 273.16 k. One kelvin is equal to $\frac{1}{273.16}$ of the triple point of water and is the same as one degree on the Celsius scale of temperature.

Kelvin's Equation. This relates the Peltier coefficient to the thermoelectric power for a conductor (or semiconductor).

$$\pi\,(B) = T\,\Sigma\,(B)$$

where π (B) is the peltier coefficient which is a function of magnetic flux density B if the conductor is placed in a magnetic field, and Σ (B) is the thermoelectric power.

Kelvin Skin Effect. A non-uniform distribution of variable currents in solid conductors, resulting in an increase in current density near the surface.

Kelvin's Statement (of the Second Law of Thermodynamics). No process is possible whose sole result is the abstraction of heat form a reservoir and the performance of an equivalent amount of work: i.e. it is impossible to construct a perfect heat engine.

Kelvin's Theorem. For an idea homogeneous fluid in a region of flow where the entropy is constant, the circulation around a closed path moving with the fluid remains constant; i.e. vorticity cannot be created or destroyed under these conditions.

Kepler's Laws. Three laws of planetary motion proposed by Johannes Kepler in 1609 and (third law) 1619. They state that: (1) each planet moves round the sun in an elliptical orbit with the sunat one focus of the ellipse; (2) the line joining a planet to the sun sweeps out equal areas in equal times, i.e. orbital veleocity decreases as distance from the sun increases; (3) the square of the *sidereal period* (P) of a planet is directly proportional to the cube of its mean distance (a) from the sun. For P in years and *a* in *astronomical* units, $P^2 = a^3$.

Kerr Effects. Two effects concerned with optical changes produced by magnetic or electric fields. In the magneto-optical effect, plane-polarized light is slightly elliptically polarized when reflected by the pole of an electromagnet. In the electro-optcal effect, the plane of polarization of a beam of light is rotated when passed through certain liquids or solids across which a potential difference is applied. This effects is utilized in the *Kerr cell,* which consists of a transparent cell containing a liquid, such as nitrobenzene; two parallel plates immersed in the liquid enable a field to be applied so that the passage of a beam of polarized light can be interruped. The cell is used as a high-speed shutter and to modulate *laser* beams. Named after the discoverer John Kerr (1824-1907).

Kerr Electro-Optic Effect. J. Kerr discovered in 1875 that certain materials, when placed in a stationary electric field, become optically anisotropic and doubly refracting. If the electric field is applied perpendicular to the direction of the light, then the effect is called the Kerr effect. If n_1 and n_2 are the refractive indices for light whose planes of polarization are respectively parallel and perpendicular to the electric field E, then

$$n_1 - n_2 = k \lambda E^2$$

where λ is the wavelength of the light and k is Kerr's constant. This effect has become of technological importance in optical communication systems, but see also the *Pockels Effect.*

Kerr Magneto-Optic Effect. Light, plane polarized in or normal to the plane of incidence, becomes elliptically polarized when reflected from the polished pole of an electromagnet.

Ketteler-Helmholtz Formula. In media which exhibit anomalous dispersion, the refractive index n for light of wavelength is given by

$$n^2 = 1 + \sum \frac{A_m \lambda^2}{\lambda^2 - \lambda_m^2}$$

if the medium contains sets of molecules with different periods of vibration. λ_m is the wavelength in free space of light of the same frequency as the natural frequency of the mth type of molecule. A_m is a constant referring to that molecule and the summation is taken to include all the natural periods of vibration of the molecules in the substance. Alternatively, it is sometimes referred to as the *Sellmeyer Formula.*

Kilogram (kg). The *SI* unit of mass equal to the mass of the platinum-iridium prototype kept at the International Bureau of Weights and measures near Paris.

Kiloton. A measure of the explosive power of a nuclear weapon. It is equivalent to an explosion of 1000 tons of trinitortoluene (TNT).

Kilowatt-hour (kW-hr). A unit of energy used in charging for electricity. It is equal to the work done by a power of 1000 watts in 1 hour.

Kinetic Energy (K.E.). The energy possessed by a body due to its being in motion. For a body of mass m moving with velocity v, the kinetic energy is given by K.E = $1/2\ mv^2$.

Kinetic, Friction. The force of friction that acts on a body when it is moving over a second body.

Kinetic Theory of Gases. A theory that explains the behaviour of gases on the assumptions that each gas is made of gas molecules in continuous motion. The theory is based on the following assumptions.

1. The gas molecules are hard and perfectly smooth elastic point particles.
2. They do not exert any significant force on one another except during collisions.
3. The volume occupied by the gas molecules is negligible as compared to the total volume occupied by the gas.
4. The duration of each collision is negligible in comparison to the interval between two successive collisions.

Kirchhoff's Equations. The relationship between the variation of heat of reaction ΔH with temeperature and the change of heat capacity ΔC_p at constant pressure p during a process is given by

$$\Delta Cp = \left[\frac{\delta(\Delta H)}{\delta T}\right]_P$$

Similarly if ΔU is the internal energy change, then the change of heat capacity ΔC_v at constant volume is given by

$$\Delta C_v = \left[\frac{\delta(\Delta U)}{\delta T}\right]_v$$

Kirchhoff's Law. Two laws applying to electrical networks, discovered by G.R. *Kirchhoff.* The first states that the net current flowing into and out of any point in the network is zero. The second states that the algebraic sum of the voltages in any closed loop of the network is equal to the algebraic sum of the products of the currents and the resistances through which they flow.

Klystron. An electronic device used to generate or amplify *microwaves.* It consists of a sealed evacuated tube in which a steady beam of electrons from an elctron gun is alternatley accelerated and retarded by high-frequency radio waves (velocity *modulation*) as it passes through a cavity. The resultant radio-frequency pulses are picked up at a second cavity, either as a voltage oscillation or, if connected to a *waveguide,* as electromagnetic waves. The second cavity can be tuned to the input frequency or a harmonic of it.

Knife-Edge. A sharp edge that is used as a fulcrum or support as in a Kater's pendulum, balance etc..The area of contact between the two parts is reduced to a minimum by use of a sharp tip.

Kirkendall Effect. If inert marker wires are embedded at the common face of a copper and a brass (CuZn) block, which is held at a hight temperature, the wires marking the original interface move into the brass. This is because the zinc atoms diffuse past the interface faster than the copper atoms. On condensing, excess vacancies can cause the formation of voids in the brass. The effect, discovered by F. C. Kikendall (1942), confirmed in the above experiment by Smigelskas and Kirkendall (1947) and since shown in many other systems, provides proof of the predominance of a vacancy mechanism for diffusion in pure metals and substitutional alloys.

Knudsen Cosine Law. Kinetic theory shows that, for a gas at rest and at a constant temperature, the number of molecules *dn* striking or leaving an area *d*A of the wall subtending a solid angle $d\phi$ at an angle θ to the normal is given by

$$dn = \frac{dA}{4\pi} \quad \bar{n}c \cos\theta \;\; d\Phi$$

where n is the number of molecules per unit volume and $\bar{c}$ is the mean velocity of the molecules.

Knudsen Flow. That flow which occurs when the mean free path of the molecules is greater than the dimensions of the apparatus. The ratio of the latter to the former is often calld *Knudsen's* number.

Knudsen Formula for Gas Flow. The mass of gas q passing any point is a tube per second at a certain temperature and very low pressure is given by

$$q = \frac{(2\pi\rho)^{1/2}}{6} d^3 \frac{p_1 - p_2}{l}$$

where $(p_1 - p_2)/l$ is the pressure gradient along the length of the tube l, d is the diameter of the tube and ρ is the density of the gas at the given temperature and unit perssure.

Kopp's Law. The psecific heat of a solid element is the same whether free or combined. In general this is not true.

Kossel Lines. These are analogous to *Kikuchi Lines* in electron diffraction and occur when a point source of *x*-radiation is used. The *x*-rays meet the atomic planes at a continuous range of angle and those striking at the Bragg angle are reflected. There will be a cone of such rays and these given rise to a black arc (Kossel line) on the photographic film, either for back reflection or transmission depending on the angle of the cone. Corresponding to the cone of diffracted rays there will be also a deficiency cone which can give rise to a white arc in the transmission pattern.

Kossel-Sommerfeld Law. The arc spectrum of an element having an odd atomic number shows even multiplicity, whereas the arc sectrum of an element having an even atomic number shows odd multiplicity.

Kramer's Theorem. In electron paramagnetic resonance in solid materials, a two-fold spin degeneracy is not removed by electric fields or spin-orbit coupling if the spin is half integral.

Kundt's Constant. A constant used to measure the *Faraday Effect*. It is similar to *Verdet's constant,* but is for rotation per unit magnetization in place of rotation per unit magnetic field.

Kudt's Rule. The refractive index of a medium on the shorter-wavelength side of an absorption band is abnormally low and on the longer-wavelength side abnormally high. The pehnomenon of anomalous dispersion is due to this effect.

Kundt's Tube. Refers to a glass tube at one end of which is fitted a light adjustable piston. The other end has been closed by a cap through which passes a metal rod clamped at its centre. A small quantity of lycopodium powder is spread uniformly throughout the tube. The free end of the rod gets rubbed to and for along its length. Stationary waves are found in the air column in the tube. A measurement of the wavelength of these wave enables one for calculating the velocity of sound in air.

Kurie Plot (for Beta Decay). In 1934 E. Fermi derived, for the energy distribution of beta particles emitted from radioactive isotopes, a formula of the form

$$\left[\frac{N(p)}{p^2 F(Z,P)}\right] = C(E_{max} - E)$$

where $N(p)$ is the number of particles having momentum between p and $p + dp$, F, (Z, p) is the correction factor for the coulombic interactions between the nuclear particles and the beta particles. E is the energy of the beta particles, which can be emitted with maximum energy $E_{max,}$ C is a constant and Z is the atomic number of the product nucleus.

The distribution can be plotted in 'straight-line' form when the graph is called a Kurie plot or *Fermi plot.*

L

Lambert. A unit of luminance equal to the luminance of a surface that emits one lumen per square centimetre. Named after J.H. *Lambert.*

Lambert's Law. The illuminance of a surface that is illuminated by a point source of light normally is proportional to $\frac{1}{r^2}$ where r is the distance between the source and the surface. If the incident rays make an angle θ with the normal to the ray, the illuminance is proportional to cos θ.

Lamb Shift (or Lamb-Rutherford Shift). Spectroscopists found a separation between the main components of the H_α line in the hydrogen spectrum to be less than predicted by the theory of Dirac. This was shown by Lamb, using quantum electrodynamic theory, to be due to $2S_{1/2}$ term (quantum numbers $l=1, j=1/2$) being 0.2836 m^{-1} higher than the $2P_{1/2}$ term (quantum numbers $l=0, j=1/2$), this difference being called the Lamb shift.

Lame's Elastic Constants. The elastic properties of an isotropic material are completely determined by two constants called Lame's elastic constants and denoted by μ and λ.μ is the shear modulus or rigidity coefficient (also denoted G). λ is not a named modulus but is related to the bulk modulus, K, by

$$K = \lambda + \frac{2}{3}\mu$$

Laminar Flow. Fluid flow in which the particles move in parallel layers, as in a fluid moving slowly along a horizontal straight pipe. Above a velocity given by the *Reynolds* number, the layers no longer remain parallel and the flow becomes turbulent.

Landau Diamagnetism. This is the net non-vanishing magnetization which is produced antiparallel to an applied magnetic field as a result of coupling of the magnetic field with orbital electronic motion. For free electrons, the diamagnetic susceptibility has one third the magnitude of the paramagnetic *Pauli Spin* Susceptibility.

Landau Fluctuations. Experimental determination of the energy loss of a a fast particle frequently involves measurement of the energy lost in a thin detector in which the particles suffer many ionizing collisions but do not lose much of their innitial energy. There are marked fluctuations in the observed rate of energy loss. This variation may be attributed to the *Poisson Distribution* of ionizing events and to the wide range of energies which can be lost by fast particles in an inelastic collision.

Landau Levels. When a magnetic field is applied to a metal at low temperatures, the electrons are forced to coalesce into quantized levels of energy $(n + \gamma)\, h\, \omega_c$ n is an integer, γ takes the value 0.5 if free electrons are assumed and ω_c is the cyclotron frequency corresponding to the magnetic field. These energy levels are called Landau levels.

Lande Splitting Factor. Lande, partly empirically, derived a formula for *g*, the fraction by which the magnetic moment of an atom must be multiplied to allow for spin, in terms of the three quantum vectors; *j* the total momentum vector, L the azimuthal quantum vector and S the moment of electron spin:

$$g = 1 + \frac{J(J+1) + S(S+1) - L(L+1)}{2J(J+1)}$$

Landolt Fringe. If a brilliant source of light is viewed through two Nicol prisms oriented with their principal axes perpendicular to one another, the field is not quite dark when exact adjustment is made. The darkened field is crossed by a black fringe, which changes, position of the prisms are rotated slightly. Lippich showed that this fringe was due to the fact that the directions of vibrations in various parts of the field were not parallel.

Langevin Diamagnet Equation. The diamagnetic susceptibility per gram-molecule for a diamagnetic substance is given by

$$Xm = -\frac{\mu_o Ze^2 N_A}{6m} < r^2 >$$

where N_A is Avogadro's number, Z the atomic number and $<r^2>$ the mean square distance of the electrons from the nucleus, assuming spherical symmetry.

Langmuir Effect. An ionization produced when atoms of low ionization potential come into contact with hot metal of high work function.

Laplace Equation. It is a linear differential equation of the second order.

$$\frac{\delta^2 V}{\delta x^2} + \frac{\delta^2 V}{\delta y^2} + \frac{\delta^2 V}{\delta z^2} = 0$$

V may, for example, be the potential at any point in an electric field where there is no free charge.

Laporte's Rule. For dipole radiation, only transitions between terms of opposite parity are allowed: i.e. transitions between terms having either $\Sigma_i l_i$ even or having $\Sigma_i l_i$ odd are allowed, where $\Sigma_i l_i$ is the summation of the absolute value of the quantum number l for all the electrons.

Larmor Precession. The action of a magnetic field on the orbital motion of an electron was stated by Larmor. He showed that a uniform field does not affect the form and inclination of the orbit to the magnetic lines of force, or the frequency and velocity of the electron in its orbit. The orbit itself, however, rotates (precesses) in a uniform fashion about an axis parallel to the field and the frequency of precession is called the *Larmor frequency.*

Laser (Light Amplification by Stimulated Emission of Radiation). A device that produces a beam of high-intensity coherent monochromatic radiation (light, infrared, or ultra violet). Stimulated emission is the emission of a photon when an atomic electron falls from a higher energy level to a lower level as a result of being stimulated by another photon of the same frequency. In the laser large numbers of electrons are "pumped" into a higher energy level, an effect called population inversion, and then stimulated to produce a high-intensity beam. Laser beams have been produced from solids, liquids and gases. The simplest type is the ruby laser, consisting of a cylinder of ruby, silvered at one end and partially silvered at the other. A flash lamp is used to excite chromium ions in the ruby to a high energy level. When the ions fall back to their ground state photons (wavelength 694.3 nanometres) are emitted. These photons collide with other excited ions producing radiation of the same wavelength (monochromatic) and the same phase (coherent), which is reflected up and down the ruby crystal and emerges as a narrow beam from the partially silvered end. Lasers are used in civil engineering, in surgery, in laser interferometers to measure very small displacements, in *holography*, and in *compact* disc players.

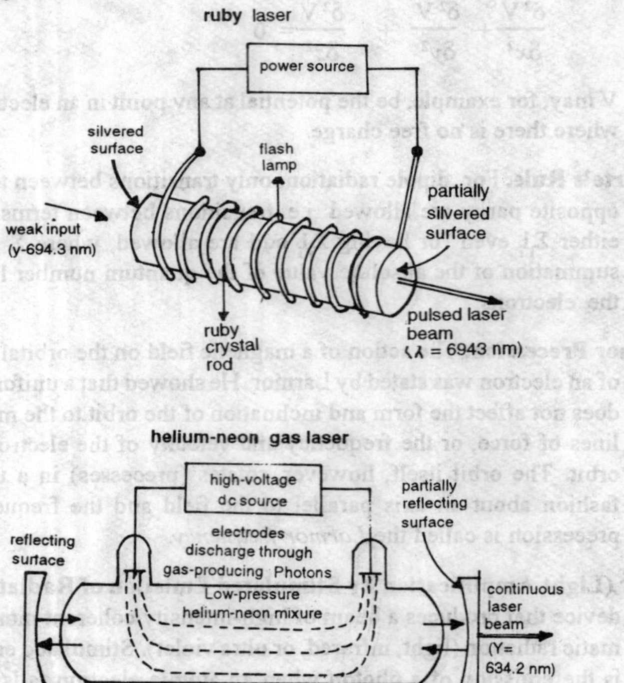

Laser. **In the solid-state ruby laser, chromium ions are excited by an intense flash of light and then stimulated by weak light of one wavelength to emit a pulse of photons. In the helium-neon gas laser, a continuous laser beam is produced from an electrical discharge through low-pressure gas.**

Laser Printer. A type of printer used with computers, in which the image is formed by scanning a photoconductive drum with a narrow-beam laser. Parts of the surface are charged and, as in a *photocopying* machine, the image is developed by applying powdered pigment, which adheres to the charged regions. The pigment is transferred to paper and fixed by heat. Laser printers are fast, quiet, and have high resolution, allowing near-typographical print quality in many cases.

Latent Heat. The quantity of heat required to change a unit mass of substance from one states without any change of temperature. Latent heat of ice, is the heat required to convert 1 gm of ice into water at 340^0 C. Its value is 80 cal/gm. Latent heat of steam = 540 cal/gm.

Lateral Inversion. Refers to the type of inversion produced in the image formed by a plane mirror. The left hand side appears as right hand side and vice versa.

Lattice Constant. The length of edge of the angle between the axes of the unit cell of a crystal. It is generally the ege length of a cubic unit cell.

Laue Diagram. When a heterogeneous beam of X-rays, electrons, or neutrons is allowed on to a thin crystal, the different atomic planes give rise to a series of symmetrically arranged spots on a photographic plate placed behind the crystal. This pattern, known as a Laue diagram, is used to determine the type of crystal structure.

Laue Equations. When diffraction of X-rays is produced by a three-dimensional array of atoms, the condition for scattering in phase can be broken down into three equations called the Laue equations which must be satisfied simultaneously. If there are three crystal axes x, y and z., and the separations of the atoms along the three axes are *a, b* and *c* respectively, then the equations are

$$a(\cos\alpha - \cos\alpha_0) = h\lambda$$

$$b(\cos\beta - \cos\beta_0) = k\lambda$$

$$c(\cos\gamma - \cos\gamma_0) = l\lambda$$

Angles α_0, β_0 and γ_0 are the angles which the incident beam makes with the *x, y* and *z* axes respectively and α, β, and γ are the corresponding angles for the diffracted beam λ is the wavelength of the radiation and *h, k* and *l* are integers coresponding to the *Miller Indices* of the planes considered to be diffracting the x-rays.

Lauritsen Electroscope. An electroscope in which a metal wire carries a metalcoated quartz fibre, that is attached to the wire when under charge. A T-piece on the fibre is seen by a microscope with an eyepiece scale.

Lawson Criterion. The product of the particle density of plasma (in particles per cm^3) and the containment time (in seconds) at or above its ignition temperature, such that the fusion energy given out equals the energy required to form and confine the plasma. For the deuterium-tritium the reaction value of the Lawson criterion is 10^{14} s/cm^3.

Leakage. Refers to the flow of an electric current, on account of imperfect insulation, in a path other than that intended. The current is small compared to that of a short circuit.

Leakage Flux. In any electrical machine or transformer, having a magnetic circuit. The flux is outside the useful portion of the flux circuit.

Least Action Principle. In a dynamic system, the motion between two points is stationary for the actual path in comparison with various paths between the same points for which the total energy has the same constant.

Least Squares, Method of. A technique for determining the equation which gives the best result for experimental data. The best result is considered to occur for a value of a at which the sum of the squares of these deviations is a minimum.

Le Chatelier's Rule (or Principle). When a constraint is applied to a system in equilibrium, a change takes place within the system, that opposes the constraints and tends to restore equilibrium.

Leclanche Cell. An *electric cell* invented in 1867 by the French engineer Georges Leclanche (1839-82). It consists of a carbon anode surrounded by manganese dioxide and crushed carbon held in a porous bag or pot dipping into a solution of ammonium chloride, which also contains a zinc cathode. It delivers 1.5 volts. The common *dry cell* consists of a Leclanche cell or a *battery* of cells in which the electrolyte is a paste of ammonium chloride in a zinc casing.

Lee's Disc. An instrument used for measuring the thermal conductivity of poor conductors. It consists of a thin flat disc of the substance to be tested, that is sandwiched between two discs of a good conducting substance continuously such as brass. The upper disc is heated to fixed temperature and the temperature of the upper and lower disc is observed. Form this, the amout of heat floweing through the simple disc and therefore its thermal conductivity can be calculated.

Leidenfrost's Phenomenon. If liquid is dropped on to a hot surface, there is a critical temperature above which wetting by the liquid does not occur. The liquid becomes insulated from the surface by a layer of vapour. As a result, water, for example, when dropped on to a hot plane, breaks into globules which move erratically across the surface.

Lennard-Jones Potential. The total potential energy $\phi(r)$ of two atoms of an inert gas at separation r is

$$\phi(r) = \frac{B}{r^{12}} - \frac{C}{r^{6}}$$

where B and C are empirical parameters. The r^{-6} term represents the long-range attractive force (see *Van der Waals—London Interaction*) and the r^{12} term the short range repulsive force.

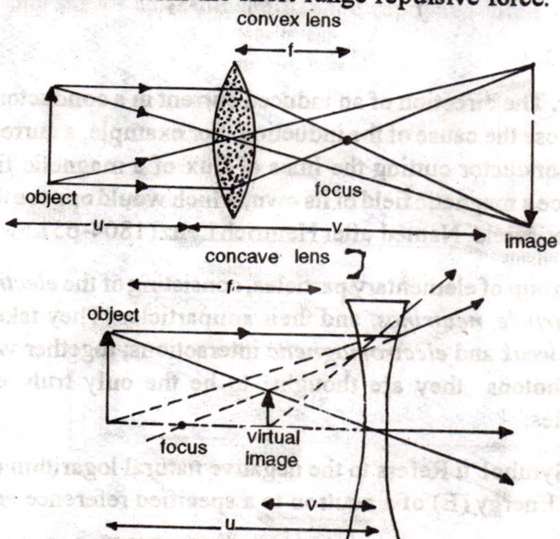

Lens. The lines representing light rays show how a convex lens gives a real inverted image and how a concave lens gives an upright virtual image.

Lens. A piece of transparent material, usually glass, quartz, or plastic, used for directing and focussing beams of light. The surfaces of a lens have a constant curvature; if both sides curve outwards at the middle the lens is called convex, if they curve inwards it is concave. The image formed by a lens may be real, in which case the rays converge to the image point (a converging lens), or virtual, in which the rays diverge from the image point (a diverging lens). The focal length of a lens is the distance from the lens at which a parallel beam of light is brought to a focus. If the focal length of the lens is *f*, the rays from an object distance *u* from the lens are focused ata disance v from the lens, where $1/f = 1/u + 1/v$ and *u, v,* and *f* all obey certain sign conventions.

Lens Waker's Formula. Refers to a relation between the focal length of a lens (*f*), refractive index of the material (μ) of which it is made and radii of curvature R_1 ,R_2 of its surfaces.

$$\frac{1}{f} = (\mu - 1)\left(\frac{1}{R_1} + \frac{1}{R_2}\right)$$

Lenz's Law. The direction of an induced current in a conductor is such as to oppose the cause of the induction. For example, a current induced by a conductor cutting the lines of flux of a magnetic field would produce a magnetic field of its own, which would oppose the original magnetic field. Named after Heinrich Lenz (1804-65).

Lepton. A group of elementary particles, consisting of the *electron, muon, tau particle, neutrinos,* and their antiparticles. They take part only in the *weak* and *electromagnetic* interactions; together with *quarks* and photons they are thought to be the only truly elementary particles.

Lethargy. Symbol. u Refers to the negative natural logarithm of the ratio of the Energy (E) of a neutron to a specified reference energy (E_0).

Lever. A rigid bar which can be turned about a fixed point, known as fulcrum. Depending upon the position in the middle, of the fulcrum, the effort or the load, the levers are classified into various kinds.

Lewis-Randall Rule. At a given temperature the fugacity of a constituent of a mixture of gases is proportional to its mole fraction. Up to 25 atmospheres this rule appears to hole, and is a great, improvement on *Dalton's Law of Partial Pressures.*

Lewis Number. Symbol: (Le). A dimensionless number used in problem that involves both heat and mass transfer. It is equal to λ /p Dc,λ is the thermal conductivity, p is the density, D is the diffusion coefficient, and *c* is the specific heat capacity.

Leyden Jar. A form of capacitor made up of a glass jar having metal foil on the lower outside an inside surfaces.

Lichtenberg Figure. A complex symmetrical star-shaped pattern which is formed by dust particles adhering to a solid dielectric when it is under a high potential. It can also be produced on a photographic plate placed between electrodes.

Lift Coefficient. For a body and a fluid of density *p* in a motion having relativ velocity V, the lift coefficient of the body is the ratio of the component of resistance for perpendicular to the direction of relative motion (lift) to the quantity $pl^2 V^2$, where *l* is some characteristics length of the body. The coefficient is dependent on the circulation round the body.

Light. The form of *electromagnetic* radiation to which the eye is sensitive. It forms the part of the electromagnetic spectrum from 740 nanometres (red light) to 400 nanometres (blue light), white light consisting of a mixture of all the colours of the visible spectrum. The nature of light has been in dispute from earliest times, *Newton* supporting a corpuscular theory in which a luminous body was believed to emit particles of light. This theory adequately explained reflection and geometric optics but failed to explain *interference* and *polarized* light. The wave theory, supported in the 19th century by *Fresnel* and *Foucault,* adequately explains these phenomena and achieved a mathematical basis when *Maxwell* showed that light is a form of electomagnetic radiation. The wave theory, however, does not explain the *photoelectric* effect and *Einstein* reverted to a form of the corpuscular theory in using the *quantum* theory to postulate that in some cases light is best regarded as consisting of energy quanta called photons. The present view, expressing *Bohr's* concept of complementarity, is that both electromagnetic theory and quantum theory are needed to explain this phenomenon.Light travels at a speed of 2.997 925 x 10^8m per second in free space, this being, according to the special *relativity,* the highest attainable velocity in the universe.

Light Visible. Electromagnetic radiation that can be seen by human eye. Its wave length lies in the range 4000-7000 A°. different wave lengths corresponds to different colours. Wave lengths below 4000 A° are known as ultraviolet light while those above 7000 A ° are known as infrared radiations. Light is produced by surfaces when they are heated to very high temperatures. The origin of light is now understood in terms of atomic theory as due to transitions between electronic energy levels in atoms and molecules of the substances on being heated.

Lightning. An electrical discharge in the atmosphere caused by the build up of electrical charges in a cloud by such methods as friction between the particles in the cloud. The potential difference causing the discharge may be as high as one thousand million volts. The electricity then discharges itself in a lightning flash, which may be between the cloud and the ground or, much more commonly, between two clouds or parts of a cloud. Thunder is the noise made by the discharge. *See also* ball lighting; thunderstorm.

Lightning Conductor. An earthed conducting rod placed at the top of buildings, etc., to protect them from damage by the lightning. It acts by providing a low-resistance path to earth for the lightning current.

Lightship. A vessel, anchored at sea, used as a lighthouse. Once in wide use most lightships have now been replaced by fixed structures or, because of advances in sophisticated navigation devices, have been eliminated entirely.

Light Year. A unit of distance, used in astronomy, equal to the distance travelled by light in one year. 1 light-year = 9.46×10^{15} metres or 5.88×10^{12} miles.

Lightning Flash. A complete lightning discharge alone a single discharge path. It may be made up of more than one lightning stroke, in which case it is known as multiple stroke.

Lightning Stroke. A discharge which is component of a complete lightning flash. It is the discharge of one of the charged regions of a thunder cloud. The polarity of a lightning stroke is the polarity of the electric charge which is brought to earth. A lightning stroke to any part of a power of communication system is known as *direct stroke.* It is known as *indirect stroke* if it induces a voltage in that system without actually striking it.

Limiter. An instrument which automatically prevents some output charaxteristic from exceeding a certain value. e.g., in a frequency modulation receiver it is used for elimination or reducing any amplitude modulation of the received signal.

Limiting Current. *Syn.* Diffusion current. In electrolysis the rate of diffusion of ions towards the electrodes cannot keep pace with their deposition. Thus there is a limiting value of current which can be passed under the particular conditions of ionic concentration and characteristics of the electrolyte.

Linde Process. A process used for the liquefaction of air. It is based on the Joule-Thompson effect. Air is compressed and expanded by passing it through a nozzle. The cooler air is used to cool the incoming compressed gas. The process is repeated many times when the temperature of the air falls, sufficiently to liquefy it.

Linde's Rule. An expression for the increase in resistivity, per atomic per cent impurity, for susbstitutional impurities introduced into a monovalent metal:

$$\Delta\rho = a + bz^2$$

where a and b are constants for a given row of the periodic table and a given solvent metal and z is the excess valency (i.e. the impurity has valency $z + 1$).

Linear Accelerator. An *accelerator* in which charged elementary particles are repeatedly accelerated along a long straight tube by a radio-frequency electric field. In modern linear accelerators the field is supplied by the electric component of a travelling radio wave in a waveguide. The particles are confined in the tube by a series of magnetic lenses, which focus the beam. The maximum energy attained by a linear accelerator is about 10 GeV for electrons and 2 GeV for protons.

Linear Expansion. The change in length of a substance when heated. Coefficient of linear expansion is defined as the change in length per unit length per degree rise of temperature and measured in per degree kelvin.

Linear Motor. A form of electric induction motor in which the stator and the rotor are linear instead of cylindrical and parallel instead of coaxial. The development of linear motors as a method of traction for monorail intercity trains has been proposed by E.R.Laithwaite (1921-89). In this arrangement one winding would be in the train and the other on the single rail, thus obviating the need for rotating parts.

Line of Force (Electric). An imaginary line in an electric or magnetic field, the tangent to which at any point gives the direction of the electric field at that point. Lines of force never cross each other.

Liquefaction of Gases. A gas can be liquefied only by the application of external pressure on it if its temperature is below its critical temperature. The gas can be liquefied by cooling it using Joule Thompson effect, by evaporating under reduced pressure and by making the gas expand against an external pressure.

Lissajous Figures. Refers to the loci of the resultant displacement of a point subjected to two or more simple harmonic motions simultaneously. If the two periodic motion are of the same frequency and are at right angles to each other, the resulting figure change from a straight line to an ellipse depending on the phase difference between the two motions.

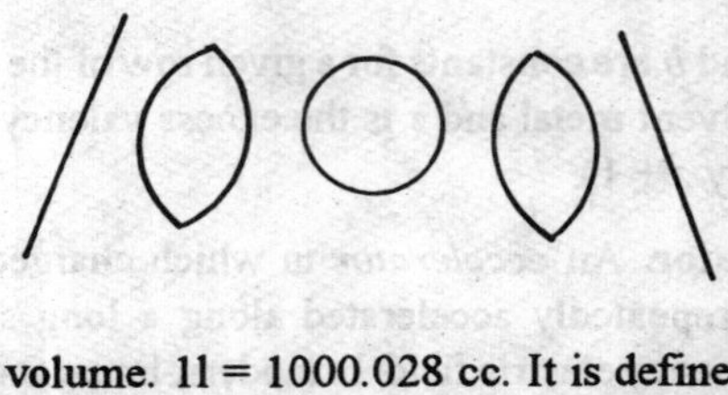

Litre (l). Unit of volume. 1l = 1000.028 cc. It is defined as the volume of 1 kg of pure water at 760 mm pressure and 4° C.

Lloyd's Mirror. An optical arrangement to study interference of light. Light incident on a mirror at grazing incidence afte reflection is made to interfere with direct light from the source. The arrangement is essentially equivalent to two coherent sources that produce interference fringes.

Load (Electrical). (1) A device or material wherein signal power is dissipated or received. (2) Refers to the power delivered by a machine, generator, or electronic circuit or device.

Loaded Concrete. Concrete, made up of material of high atomic number or capture cross section, such as barium, iron, or lead, to increase its effectiveness as a radiation shield in nuclear reactors.

Load Impedance. Refers to the impedance presented by a load to the driver circuit.

Load Line. A line drawn on the graph of a family of characteristics, that shows graphical relationship between voltage and current for the particular load of the circuit under consideration.

Locomotive. An engine that draws a train on a *railway*. The first locomtives, designed by *Trevithick* and *Stephenson* were driven by steam engines and steam dominated the railways until the end of World War II. Even in the mid-1970s, it was estimated that over 25 000 steam locomotives were still in use throughout the world. However, the steam engine has a low efficiency (about 8% in a locomotive, which does not use a condenser), it takes a long time to become operational while stream is raised, it uses an awkward and dirty solid fuel (which it has to pull with it in a tender immediately behind the engine), and it creates pollution. For these reasons steam locomotives have largely been replaced in the industrial countries by electric, Diesel-electric, or Diesel trains, all of which have efficiencies of about 22%.

Where the traffic justifies the cost of installing ovehead wires or a conductor rail, electric trains are usually preferred, as it is more efficient to generate electricity centrally than in the locomotive. Various configurations have been tried, but series-wound direct-current motors are the most widely used, with rectification in the locomotive when the supply is alternating current. In some cases a three-phase supply is used, with *thyristor* control. For lines in which a permanent installation is not economic, Diesel engines are used as a prime mover. These are either coupled hydraulically to the wheels, using a hydraulic torque converter (or fluid flywheel) and a gearbox, or to an electric generator or alternator, which produces current to power electric motors that drive the wheels. The first Diesel-electric train was used in Sweden in 1912 and the first Diesel-hydraulic in Germany a year later. Because the Diesel provides a very low torque at low speeds, it cannot be used without a hydraulic coupling, but even these units are usually restricted to small trains.

The use of *gas* turbines on the railways began in 1934, when a turbine powered a Swedish experimental train. A gas turbine-electric train was first tried in Switzerland in 1941. Although the gas turbine is lighter than the Diesel, it is only efficient at high power, and most of the long runs that could utilize its smooth full-power output are already electrified. The outlook for gas turbines is therefore not regarded as very promising.

Longitudinal Wave. A wave wherein the particle displacement is in the same direction in which the wave travels. e.g. sound wave.

Lorentz Force. The force F acting on a moving charge q in magnetic and electric fields is given in SI by

$$F = q\,(E + v \times B)$$

where v is the velocity of the charge, E the electric field and B the magnetic induction. Often only the magnetic term in the force is referred to as the Lorentz force. The force can then be expressed into non-vectorial form

$$F = qvB \sin\theta.$$

θ is the angle between the direction of motion of the charge and the direction of the magnetic field, and the force is perpendicular to both the direction of motion and the field.

Lorentz Gauge. For an electromagnetic field, knowledge of the electric and magnetic fields is not sufficient to determine uniquely the scalar potential ϕ and the vector potential A of the field. Hence an auxiliary condition is chosen (the gauge) giving the equations a symmetrical form. The Lorentz gauge is the condition (in SI units):

$$\Delta . A + \frac{1}{c^2}\frac{\delta\phi}{\delta t} = 0.$$

Lorentz-Lorenz Law

$$\frac{n2-1}{n^2+2} = \frac{4\pi}{3}\ \Sigma\alpha$$

where n is the refractive index and $\Sigma\alpha$ is the sum of polarizabilities of all particles present in unit volume. α is defined by

$$p = \alpha E$$

where p is the electric dipole induced by an electric field E.

Lorentz Polarization Factor. In the scattering of X-rays by the atoms of a single crystal, the Lorentz polarization factor LP is given by

$$LP = \frac{1 + \cos^2 2\theta}{2 \sin 2\theta}$$

where θ is the Bragg angle. Tables of this factor versus θ are available.

A similar Lorentz polarization factor is available tabulated as a function of θ for scattering from a powder sample, as in the *Debye-Scherrer Method.* Here

$$LP = \frac{1 + \cos^2 2\theta}{\sin^2\theta\ \cos\theta}$$

Lorentz's Theory of Light Sources. When Maxwell showed that light was an electromagnetic wave, it was necessary to identify some electrical oscillator of atomic proportions to account for these waves.

Lorentz considered that electrons which are normally held in an equilibrium position are capable of vibration when displaced under the influence of a restoring force. The bound electron was supposed to be the source of the electromagnetic waves of light when executing damped vibrations. It was also capable of absorbing light when the frequencies of the vibration and of the light were in agreement.

Lossy. Relating to a material or instrument such as a dielectric or transmission line, which dissipates energy with a high degree of attenuation.

Loudness of Sound. The sensation produced by a source of sound in a listner. It depends on the intensity of sound for a given frequency. It also depends on the frequency of the sound.

Loudspeaker. A device for converting electrical signals into sound. It usually consists of a small coil fixed to the centre of a movable diaphragm or cone. The coil is in an annular gap between the poles of a strong *magnet.* An audio frequency electrical signal fed to the coil creates a varying magnetic field, which interacts with the steady field in the gap. This causes the coil, and the attached cone, to vibrate and produce sound waves of the same frequencies as the electrical signal. Generally, larger cones give a better response at low frequencies and the smaller cones are best at hight frequencie; for the best results two or more different-sized cones are therefore used, either in the same or in separate cabinets. The loudspeaker cabinet is also an important part of the system since it can act as a sound baffle and improve the frequency response.

Low Frequency. Frequency in the range of 30-300 Hz.

Lumen (lm). The *SI* unit of luminous flux equal to the light emitted per second in a cone of one steradian solid angle by a point source of one candela.

Luminance. The *luminous* intensity of a surface in a given direction per unit of orthogonally projected area of that surface. It is measured in candela per square metre.

Luminescence. The emission of light radiation by a body when an electron of the atoms jumps from an excited state to the ground state. It is known *Phosphoroscence* if the luminescence remains even after the source of excitation of the atom is removed. Otherwise it is known *fluoroscence.*

Luminous Energy. Symbol: Q_v, Q. The product of luminous flux and its duration : $Q_v = \int \Phi v \, dt$. Measured in lumen-second.

Luminous Exitance. Symbol: M_v, M. The luminous flux that leaves a surface per unit area. Previously called the *luminous emittance.* Measured in lumen per square meter.

Luminosity. The intrinsic brightness of an object, such as a star, equal to the total energy radiated per second from the object. A star's luminosity increases both with surface temperature and with surface area: the hotter and larger a star, the greater its luminosity. Stellar luminosity is related (logarithmically) to absolute *magnitude.*

Luminous Flux. The rate of flow of light energy, taking into account the sensitivity of the observer or detector to the different wavelengths. For example, the human eye is most sensitive to the colour green. Luminous flux is measured in *lumens.*

Luminous Intensity. The amount of light emitted per second by a point source per unit solid angle in a specified direction. It is measured in *candela.*

Lux (lx). The SI unit of intensity of illumination equal to the illumination resulting from a flux of one lumen falling on an area of one square metre. Former name: metre candle.

Lyddane-Sachs-Teller Relation. In an infinite ionic crystal, electromagnetic waves do not propagate in a frequency range $\omega_T < \omega < \omega_L$ where ω is the frequency of the electric field. ω_L is also the longitudinal optical phonon frequency for the phonon wave-vector *k* tending to zero and ω_T the transverse optical phonon frequency for a large wave vector. Then, if $\varepsilon_r(0)$ is the static dielectric constant of the crystal, and $\varepsilon_r(\infty)$ the dielectric constant at a frequency at which electronic polarizability is effective but no ionic polarizability, the Lyddane-Sachs-Teller relation is

$$\frac{\varepsilon_r(0)}{\varepsilon_r(\infty)} = \frac{\omega^2 L}{\omega^2 T}$$

M

Mach's Principle. The properties of space have no independent existence but depend on the mass content and distribution within it. Thus acceleration dealt with the Newtonian mechanics can only have meaning if it is an acceleration with respect to the stars or with respect to something equally well-defined.

Magnet. A body that has an appreciable external *magnetic field* (*see also* magnetism). Every magnet has two distinct areas around which the field is greatest—called the north and south poles. Like poles repel each other and opposite poles attract each other. Ferromagnetic materials (*see* ferromagnetism) are attracted to magnets because the magnet induces a field in the material in line with its own field. *Permanent magnets* are made of ferromagnetic materials and retain there magnetism unless they are heated above a certain temperature or are demagnetized by an opposing field. *Electromagnets* only function when a current flows through their coils. The field strength along the axis of the coil is proportional to the number of turns of the coil and the curent flowing through it.

Magentic Amplifier. A transductor arranged in such a way that a small controlling d. c. input can produce large changes in coupled a.c. circuits. This a.c. can be wholy or partially rectified, and if required can be fed back in a positive or negative direction, to give wide changes in the controlling conditions and output.

Magnetic Bottle. An arrangement of magnetic fields designed to contain a *Plasma*. Usually the fields are linear with strong magnetic fields called *magnetic mirrors* at both ends so that the plasma is confined within a cylinder. They are used in experimental *thermonuclear* reactors.

Magnetic Circuit. Refers to the completely closed path described by a given set of lines of magnetic flux.

Magnetic Constant (μ_0). A constant occurring in magnetic equations.

Also known as the *permeability* of free space, its value is 4π × 10-⁷ henry per metre. It is related to the *electric* constant (ε_0) by $\mu_0 \varepsilon_0 = 1/c^2$, where c is the speed of light.

Magnetic Declination. The angle between geographical north and the horizontal component of the *geomagnetic* field. It is also known as the magnetic variation.

Magnetic Dip. The angle between the horizon and a compass needle swinging in the vertical plane. It thus indicates the direction of the vertical component of the *geomagnetic* field. It is measured with a *dip circle,* a vertically mounted magnetic needle.

Magnetic Dipole. Dipole magnetic.

Magnetic Disc. A disc coated in magnetic material on which computer information is stored, using the same technology as a *tape recorder*. A *hard disc* (or *fixed disc*) comprises one or more metal discs, usually sealed in an airtight container; it is usually an integral part of its computer. Hard discs have high storage capacities—at least 10 million characters. A *floopy disc,* made from plastic is more robust and can be removed but can store only up to 1-1/2 million characters.

Magnetic Elements. These are the magnetic declination, magnetic dip and the horizontal component of Earth's magnetic field which completely define the Earth's magnetic field at any point of the Earth's surface.

Magnetic Field. The concept, devised by *Faraday,* to explain the action-at-a-distance forces produced by a *magnet*. The magnet is thought of as being surrounded by a field of force, within which its magnetic properties are effective. The strength and direction of the field is indicated by the lines of force that join the magnet's north and south poles. A wire carrying an electric current is surrounded by a magnetic field, with concentric lines of force. An electromagnet usually consists of a coil of wire, in which the lines of force run through the centre of the coil and around its circumference. The strength of the field at the centre of the coil is proportional to the current and the number of turns, and inversely proportional to the coil's radius.

Magnetic Flux. A measure of the current-inducing properties of a magnetic field. It is measured in *webers.*

Magnetic Flux Density. Syn. magnetic induction. Symbol: B. The magnetic flux that passes through unit area of a magnetic field in a direction at right angles to the magnetic force. The vector product of the magnetic flux density and the current in a conductor gives the force per unit length of conductor and is measured in teslas.

Magnetic Focussing. Focussing of charged-particle beams by making use of shaped magnetic fields. The method is made use of in accelerators, cathode ray tubres and electron microscopes.

Magnetic Indution (*also called* **Flux Density**). A vector that specifies the magnitude and direction of the magnetic field at a point. It is equal to the magnetic flux through a unit area of a magnetic field in a direction perpendicular to the direction of the magnetic force.

Magnetic Induction. The force experienced by a unit N-pole placed at a point in a magnetic field is known as magnetic intensity at that point. Its unit is ampere per metre ($A\,m^{-1}$). Also known as magnetic field strength.

Magnetic Levitation (Maglev). A magnetic method of raising a vehicle above its tracks to provide almost frictionless propulsion. Experimental maglev trains in Britain (Birmingham's National Exhibition Centre) and Japan (Miyazaki) have shown the system to be fast but expensive. Starting on normal wheels, the train levitates above its elevated track as it gathers speed as a result of the interaction between superconducting magnets buried in the track and those in the base of each coach. Forward propulsion is by *linear motor*.

Magnetic Meridian. A vertical plane which contains the direction of the Earth's magnetic field.

Magnetic Moment. A measure of the strength of a *magnet* in terms of the torque or twisting force it experiences in a uniform magnetic field. It is equal to the product of the strength of a magnet's poles and the distance between them.

Magnetic Monopole. A hypothetical particle that would carry a magnetic charge equivalent to either a north pole or a south pole. Such particles would be analogous to charged elementary particles and would provide a complete symmetry between electricity and magnetism. Unlike an electrically charged particle, a stationary magnetic monopole would give rise to a magnetic field and, when moving, an additional electric field.

Magnetic Screening. An area may be screened from magnetic effects by enclosing it with material having high permeability.

Magnetic Sheel. Refers to a thin iron sheet magnetized across its thickness. It can be considerd as an infinite number of small bar magnets.

Magnetic Shunt. A piece of magnetic material mounted near a magnet in an electrical measuring instrument, and having means so that its position relative to the magnet can be adjusted whereby the useful magnetic flux of the magnet can be changed.

Magneic Storm. A large transient variation in the *geomagnetic* field, lasting for minutes or days. It is caused by *solar* flares producing irregularities in the normal interaction between the *solar* wind and the *magnetosphere*. As a result electron densities in the **ionosphere* are disturbed, causing problems with radio communication and radar transmission, as well as induced currents in electric power lines and in oil and gas pipelines, especially in higher latitudes.

Magnetic Susceptibillity. It is ratio $\frac{M}{H}$ where M is the magnetization produced in a substance when it is placed in a magnetic field strength H.

Magnetic Tape. A plastic strip having a coating of iron oxide, used for magnetic recording or for storing binary information in the form of rows of magnetized dots, typically 7 or 9 across the tape, and 200 to 1600 per inch along it.

Magnetic Variation. Variations in the magentic elements at a point on the surface or Earth with time. Such changes may take place suddenly due to a magnetic stone or may be slow continuous changes with passage of time.

Magnetic Viscosity. There is a time lag between application of magnetic field and the resulting magnetization in most ferromagnetic substances which is accounted for by the eddy currents induced in the substance. In some materials, however, the persistence and magnitude of the change of magnetization are great to be accounted for in this way, and this phenomenon is known as magnetic viscosity.

Magnetic Well. A configuration of magnetic fields for storing a plasma in experimental fussion devices.

Magnetism. A phenomenon in which one body exerts a force on another body with which it is not incontact (action at a distance). The space in which such a force exists is called a *magnetic* field. Stationary charged particles are surrounded by an *electric* field; when these particles move or spin, an associated effect, the magnetic field, is created. Thus, an electric current, consisting of a flow of electrons, produces a magnetic field around the conductor. The behaviour of materials in such a field depends on whether the spinning electrons in its atoms align themselves to reinforce or oppose the external field.

Magnetization. Symbol: M. The difference between the ratio of the magnetic flux (B) to the magnetic constant (μ) and the magnetic field strength (H): $M = B/\mu—H$ measured in amperes per metre.

Magnetization Curves. The magnetic properties of ferromagnetic substances are generally studied by plotting curves relating the magnetization of the material to the strength and changes of the magnetizing field.

Magneto. An *electric* generator with a permanent *magnet,* rather than an electromagnet, to create the magnetic field. It consists of one or more conducting coils rotating between pairs of *magnetic* poles. The voltage frequency is equal to the number of magnets times the speed of rotation of the coils. It is used in simple petrol engines to produce ignition.

Magnetocaloric Effect. *Syn.* thermomagnetic effect. A fall in temperature that occurs when a substance suffers adiabatic demagnetization. It increases as the initial temperature of the substance is decreased so that the effect can be used for the production of temeperature in the neighbourhood of absolute zero.

Magnetodamping. Refer to an increase is internal damping of acoustic vibrations in a metal such as nikel when it is subjected to a strong magneic field.

Magnetohydrodynamics (MHD). A method of generating electricity in which current carriers in a fluid are forced by an external magnetic field to flow between electrodes placed in the fluid. Usually the fluid is a hot ionized gas or plasma in which the current carriers are electrons. The electron concentration is increased by adding substances of low ionization potential (e.g. sodium or potassium salts) to the flame. The method has been used to increase the efficiency of generation of a gas turbine, the exhaust flame of which is used for MHD generation.

Magnetometer Deflection. An instrument used for comparing magnetic moments or fields of two magnets. It is made up of a small bar magnet pivoted just above the centre of a circular scale graduat in degrees. A long light pointer is attached to the magnet at right angles to it. In a vibration magnetometer, a magnetic needle is made to vibrate in a S.H.M. and a measurement of its period enables to calculate the magnetic field strength.

Magnetomotive Force (mmf). A measure of the magnetic effect of an electric current in a coil. It is analogous to *electromotive* force and is measured in ampere-turns, being dependent on the number of turns in the coil.

Magnetosphere. A region surrounding a planet in which charged paticles are controlled by the magnetic field of the planet rather than by the interplanetary magnetic field carried by the *solar* wind; beyond a magnetosphere, solar-wind particles flow undisturbed. Its shape arises from the interaction between solar wind and planetary magnetic field. The earth's magnetosphere, which includes the *Van Allen* radiation belts, extends 60 000 km from the sunward side of the planet but is drawn out to a much greater extent on the opposite side.

Magnetostatics. It is the study of the effect of stationary magnetic poles. Even if free magnetic poles are known not to exist, the results have some validity.

Magnetostriction. The mechanical deformation of a ferromagnetic amaterial when it is subjected to a magnetic field. The effect is the result of internal mechanical stess that arises because the energy required to magnetize the crystal domains varies with their orientation in the field. The converse effect, mechanical stress causing a change in magnetization, also occurs. Magnetostriction is used in *echo-sounding* oscillators to produce the ultrasonic sound wave.

Magnetron. An electronic device used to generate and amplify *microwaves.* It consists of a sealed evacuated tube containing a central cylindrical cathode (source of electrons) inside a cylindrical anode to which electrons are drawn by an electrostatic field. A steady magnetic field applied along the axis of the tube deflects the electrons from their radial path causing them to rotate around the cathode setting up microwave-frequency oscillations. It is widely used in radar generators.

Magnification. In optical systems, the ratio of the width of an object to the width of its image, both being measured perpendicular to the axis of the system. For a single lens this reduces to the ratio of the distances of the image and the object from the lens when the image is in focus. For optical instruments the magnification is defined as the ratio of the size of the image on the retina produced by an object with and without the instrument.

Magnox. Refers to a group of proprietary magnesium alloys used to encase the fuel elements in certain types of nuclear reactors. (See gas-cooled reactor). Magnox A consists of magnesium containing 0.8% aluminium with 0.01% beryllium.

Magnus' Effect. When a rapidly spinning body is hung up in an air current moving at right angles to its axis,it is deflected at right angles to both the current and its axis and moves to the side where its peripheral motion is in the same direction as the current.

Majorana Effect. When light is passed through certain liquid or crystalline materials in a direction perpendicular to an applied magnetic field, the material becomes doubly refracting.

Majorana Force. An exchange force between nucleons in which charge and spin are exchanged (equivalent to exchange of position). If the nucleons are represented by wave-functions, then if the wave-functions are symmetric with respect to interchange of the space

coordinates the Majorana force is attractive; if the wave-functions are antisymmetric, the Majorana force is repulsive.

Major Diatomic Scale. A musical scale which is represented as follows:

C	**D**	**E**	**F**	**G**	**A**	**B**	**C**
do	**re**	**mi**	**fa**	**so**	**la**	**ti**	**do'**
sa	**re**	**ga**	**ma**	**pa**	**dha**	**ni**	**(sa)**

Intervals between successive notes: $\frac{9}{8}$, $\frac{10}{9}$, $\frac{16}{15}$, $\frac{9}{8}$, $\frac{10}{9}$, $\frac{9}{8}$, $\frac{16}{15}$

The three intervals $\frac{9}{8}$, $\frac{10}{9}$ and $\frac{16}{15}$ are known as major tone, minor tone and semi-tone respectively.

Majority Carrier. The type of carrier forming more than half of the total charge carrier concentration.

Making-current. Circuit-breaker, or similar apparatus. The peak value of the current during the first cycle after the circuits closed on a short-circuit.

Maksutov Telescope. A telescope made up of a concave spherical mirror.

Malter Effect. If a layer of semiconductor having secondary emission ratio is separated by a thin film or insulator from a metal plate, it becomes strongly positively charged on electron bombardment. The potential may be up to 100 volts with an insulating layer of about 0.1 μ *m* thickness.

Malus' Law. If a plane polarized beam of light is allowed to fall on a polarizer, the intensity of the transmitted beam is proportional to the square of the cosine of the angle between the plane of polarization of the incident light and the plane of polarization that would be required for total transmission of the beam.

Malus' Theorem. A system of rays normal to a wave front remains normal to the wave front after any number of refractions and reflections.

Mangin Mirror. Refers to a diverging meniscus lens, silvered on the

outer convex surface. A combined refraction and reflection results in system corrected for spherical aberration and coma. It is used to throw out a parallel beam form a light source or to bring parallel light to a reasonably good focus.

Manometer. An instrment for measuring pressure differences. The simplest form consists of a U-shaped tube containing mercury, one arm of which is connected to a source of pressure and the other is open. The difference in height of the liquid in the arms is a measure of the pressure difference.

Marconi Aerial. A vertical aerial one quarter of a wavelength long earthed at its lower end through an inductive capacitative reactance.

Mark Space Ratio. The ratio of the duration of the pulse to the time between pulses.

Maser (Microwave Amplification by Stimulated Emission of Aradiation). A device that works on the same principle as the *laser,* the radiation produced being in the *microwave* region instead of in the visible spectrum. Masers are used as oscillators (e.g. the *ammonia clock*) and amplifiers.

Mask. A means of shielding selected areas of the semiconductor during the various processing steps. The cirucit layout is described on a set of photographic masks that are used during photolithography processes to difine the patterns of openings in the oxide layer through which the various diffusions are made, the windows through which the metal contacts are formed, and the pattern in which the desired metal interconnections are etched.

Mason's Hygrometer. A hygrometer of the wet and dry bulb type used as the standard British insturment. The insturment must be exposed to an air draught of 1 to 1.5 metres/second.

Mass and Weight. Two physical quantities used to express the extent to which a substance is present; they are sometimes confused. The mass of a body was defined by *Newton* as the ratio of a force applied to the body to the acceleration it produces. This is now called the *inertialmass,* as it is a measure of the extent to which a body resists a change in its motion. *Gravitational mass* is defined in terms of the gravitational force between two bodies in accordance with Newton's law of gravitation. Lorant *Eotvos* showed experimentally that inertial mass and gravitational mass are equal, a result used by Einstein in his general theory of relativity. Mass was also shown by Einstein, in his special theory of relativity, to be a form of energy, according to the relationship $E = mc^2$ (where c is the speed of light). The total of the mass and the energy of a closed system remains unchanged under all circumstances (the law of the conservation of mass and energy).

Weight is proportional to gravitational mass, being the force by which an object is attracted to the earth. It is therefore equal to the product of the mass and the *acceleration* of free fall (i.e. $W = mg$). Thus, the weight of a body may vary according to its position; the mass is a constant. In scientific terms mass and weight are different, mass being expressed in units of mass (e.g. kilograms) and weight in units of force (e.g. newtons) although, of course, in everyday usage they are both measured in kilograms, pounds, etc.

Mass Defect. The difference between the mass of a nucleus and the sum of the masses of nucleus.

Massieu Function. Symbol: J. It is equal to the negative of the Helmholtz function divided by the thermodynamic temperature.

Mass-Luminosity Law. A law relating the mass m and luminosity, L of normal stars : $\log L = 3.3 \log m$, where m and L are in solar units.

Master Oscillator. An oscillator having high inherent frequency stability, used to establish the carrier frequency in radio transmission.

Materialization. The process of the direct conversion of energy E into mass m according to Einstein's equation $E = mc^2$ (c = velocity of light).

Matthias' Rules. Empirical rules for the transition temepratures of superconducting metals and alloys;

(1) The transition temperature of an element depends regularly on its position in the periodic table, but the transition metals show entirley different dependence on their valency to that exhibited by metals with completed d-shells.

(2) For alloys between transition metals, or between non-transition metals, the transition temperature of the alloys is a smooth function of the mean valency v, weighted according to the atomic fractions of the constituent metals.

(3) For elements and alloys with closed *d*-shells, the transition temperature rises form 0 K at $v = 2$ to a maximum at $v = 5$.

(4) In the transition series, the transition temperature has a maximum at $v = 5$ and $v = 7$ with a sharp minimum at $v = 6$.

Matthiessen's Rule. If the concentration of impurity atoms in a metal is small, the resistivity caused by scattering of electron waves by impurity atoms is independent of temperature. This resistivity is additive to that caused by the thermal motion of the periodic lattice of the metal, this latter resistivity being a function of temperature.

Matthiessen's Standard. A measure of the conductivity of copper expressed as a percentage of 0.1539 (if hard-drawn) of 0.1508 (if annealed ohm for a specimen of length 1 metre and of weight 1 gram.

Maxwell-Boltzmann Distribution Law. In a non-quantized system of N particles of mass *m*, the number of particles with velocities lying between *v* and *v* + *dv* is

$$dN = 4\pi N \left(\frac{m}{2\pi k_B T}\right)^{3/2} V^2 \exp\left(-\frac{mv^2}{2k_B T}\right) dv$$

Maxwell-Boltzmann Statistics. A method of studying the probabilities of the states of a system of non-quantized particles. defined by the average values of the position, velocity or energy coordinates, in a small but finite volume of the system.

Maxwell Effect (Flow Birefringence). If a liquid is viscous, consists of anisotropic molecules, and flows in such a way that a shearing velocity gradient occurs, then it exhibits optical birefringence. If the liquid has velocity v_x (i.e. velocity in the *x* direction only), then for light travelling in the *z* direction the principal planes of polarization are at 45° to the x and y-axes, except for very high velocity gradients. The difference in refractive index is given by

$$\Delta n = C\left(\frac{\delta v_x}{\delta y}\right)$$

where c is *Maxwell's constant,* which is a measure of birefringence.

Maxwell Equation (for Refractive Index). For a non-absorbing nonmagnetic material, the dielectirc constant ε_r and refractive index n are related by the equation

$$\varepsilon_r = n^2.$$

Maxwell's Demons. Spontaneous changes in temperature and pressure such as to invalidate the second law of Thermodynamics could possibly be brought about by the intervention of intelligent beings: 'Maxwell's demons'. Thus, by means of a one-way valve, they might collect more molecules in one part of a system than in another, and thereby cause an increase of pressure. In fact experience shows that such 'demons' do not exist in closed systems in temperature equilibrium.

Maxwell (Mx) C.G.S. unit of magnetic flux 1 Mx = 10^{-8} *wb*.

Maxwell's Formula. A formula that connect the relative permittivity ε_r or a medium and its refractive index n. If the medium is not ferromagnetic the formula is $\varepsilon_r = n^2$.

Maxwell's Rule. Every part of an electric circuit is acted upon by a force which tends to move it in such a direction as to enclose the maximum amount of magnetic flux.

Maxwell's Theorem. In an elastic structure, if a load applied at one point produces a given deflection at another point, then the same load applied at the point where the deflection was first measured with produce the same deflection as before at the point where the load was first applied.

Maxwell's Theory of Light. Clerk Maxwell showed in 1860 that the propagation of light could be ragarded as an electromagnetic phenomenon, the wave consisting of an advance of coupled electric and magnetic forces. If an electric field is varied periodically, a periodically varying magnetic fields is obtained which in turn, generates a varying electric field; thus the disturbance is passed on in the form of a wave.

Maxwell's Thermodynamic Equation. For a thermodynamic system,

$$\left(\frac{\delta p}{\delta T}\right)_V = \left(\frac{\delta S}{\delta V}\right)_T \qquad \left(\frac{\delta V}{\delta T}\right)_P = -\left(\frac{\delta S}{\delta p}\right)_T$$

$$\left(\frac{\delta p}{\delta S}\right)_V = -\left(\frac{\delta T}{\delta V}\right)_S \qquad \left(\frac{\delta V}{\delta S}\right)_P = \left(\frac{\delta T}{\delta p}\right)_S$$

where p is the pressure, V the volume, T the absolute temperature and S the entropy.

Mean Free Path. The average distance covered by a particle between collisions. For a gas, it covered inversely proportional to the pressure of density of the gas.

Machanical Advantage. The ratio of the force output of a machine to the force input, i.e., the ratio of load to effort. It is useful only as an analysis of simple machines, such as levers, pulleys, jacks, etc., as no account of friction is taken. The *velocity ratio* of the machine is the distance moved by the effort divided by the distance moved by the load; the *mechanical efficiency* of a machine is the ratio of its mechanical advantage to its velocity ratio.

Mechanical Engineering. The branch of engineering concerned with the application of scientific knowledge to dynamical structures and systems, rather than the static structures of civil engineering. This branch encompasses, the design, manufacture, and maintenance of machines of all kinds, engines, vehicles, and many aspects of industrial manufacturing. The subject has numerous specialized subdivisions, including *aeronautics*, motor engineering, machine-tool design, etc. In the UK the professional body is the Institute of Mechanical Engineers.

Mechanical Equivalent of Heat. Joule's equivalent.

Mechanics. The study of the motion of bodies and systems and the forces acting on them. The subject is traditionally divided into statics, the study of bodies in equlibrium, and dynamics, the study of forces that affect the motion of bodies. Dynamics is further divided into kinetics, the effects of forces and their moments on motion, and kinematics, the study of velocity, acceleration, etc., without regard to the forces causing them. Aristotelian mechanics was based on the erroneous concept that a force is required to maintain motion. *Newtonian* mechanics recognizes that once a body is moving a force is required to stop it but no force is needed to keep it moving. Newtonian mechanics is the mechanics of classical systems, i.e. large-scale systems moving at relatively low velocities. The more general relativistic mechanics is also applicable to systems moving at speeds comparable to that of light (it reduces to Newtonian mechanics at velocities that are small compared to that of light). *Fluid* mechanics is the application of mechanical principles to fluids, both stationary (hydrostatics) and flowing (hydrodynamics).

Mechanomotive Force. The root-mean-square value of the force developed in any machine developing an alternating force.

Medical Physics. Refers to the application of physics to medicine, as in radiotheraphy, nuclear medicine, and medical electronics.

Medium Frequency. A radio frequency in the range 3×10^5 to 3×10^6 Hz.

Mega (M). A prefix meaning a multiple of 10^6.

Meissner Effect. A bulk superconductor in a weak magnetic field exhibits perfect diamagnetism. Thus, if a specimen in the normal state is cooled through the transition temperature for superconductivity, the magnetic flux originally present in the sample is expelled, except for slight penetration at the surface.

Melde's Experiment. An experiment for verification of laws of transverse vibration of strings.

Meniscus. It is the curved surface of a liquid or a column of a liquid in a tube or vessel. The shape of the meniscus depends on the angle of contact between the surface and the liquid. If the angle 90^0, the meniscus is concave. If it > 90^0, its convex.

Mercury Cell. A primary cell made up of a zinc anode, a mercuric oxide cathode and KOH is electrolyte. Its *e.m.f.* is 1.3 *v*.

Mercury Thermometer. A thermometer that uses mercury (in glass) as a working substance. It is made up of a thin capillary tube graduated in degrees and having a bulb containing mercury at its base. When the bulb is contained in a bath whose temperature is to be measured, the extent to which mercury expands is a measure of the temperature of the bath which can be read if from the scales.

Mesons. A group of unstable elementary particles (lifetimes between 10^{-8} and 10^{-15} second) that are classified a hadrons; each meson is believed to consist of a quark anti-quark pair.

Metal Detector. A device for detecting the presence of metal, usually below ground. It is sued to trace pipework and by archaeologists and other seeking buried metal. Developed from wartime mine detctors, they usually consist of an electronic circuit that oscillates at a frequency controlled by an induction coil. The inductance of the coil, which is contained in the search head of the device, is altered in the pressence of metal. The resulting change in frequency creates an audible tone in headphones or causes a light to flash.

Metallizing. The covering of an insulating material with a film or metal or other substance so as to render it electnically conducting. The technique is used in solid-state electronics. The conducting film is etched and forms interconnections on integrated circuits. It is also used in forming bonding pads for integrated circuits and discrete components.

Metaphysics. The study of existence or being in general. The term derives from the title given to a group of Aristotle's writings by the philosopher Andronicus of Rhodes (1st century BC). The status of metaphysics has been much debated; Kant thought this kind of investigation impossible because our minds can only cope with the phenomenal world or the world of appearances and *Ayer* used the word in *Language, Truth and Logic* (1936) as a pejorative term to indicate the meaninglessness of much traditional philosophy.

However, with *ethics* and *epistemology,* metaphysics is still held to be one of the main divisions of philosophy.

Metastable State. Refers to a state of a system in which it is apparently in a stable equilibrium; however, if slightly disturbed the sytem changes to a new state of lower energy.

Meteor. A lump of matter from space that enters the earth's atmosphere and is detected either optically on account of its luminosity or by ratio means on account of the trial of ionized gas left in its wake.

Meteorograph. An instrument for recording some or all of the following: temperature, relative humidity, pressure, and wind force. It is generally carried into the upper air by a balloon or kite.

Meteorology. The science of the atmosphere, particularly with respect to weather and climate.

Metre Bridge. An arrangement having a wire of one metre length and suitable terminal points so that it can be used as a Wheatstone bridge by connecting suitable resistance across the terminals. The wire porivides resistors of the two arms of a conventional wheatstone bridge that can be changed by sliding a jockey across it.

Metric System A system of measurement based on the decimal system. First suggested in 1585 by Simon Stevin (1548-1620), an inspector of dykes in the Low Countries, it was not given formal acceptance until 1795. when a French Law provided definitions for the *metre, stere, litre* and *gram*. In 1799 a subseuent law established legal standards, made of platinum, for the metre and the kilogram. However, this form of the metric system was not widely used, even in France, until the 1830s; during this period it was also adopted by most European countries.

Although the metric system was authorized in the UK in 1864, bills for its compulsory adoption were defeated in parliament in 1871 and again in 1907. British reluctance to abandon its own "Imperial units" did not weaken until 1963, when the Imperial yard and pound were given legal definitions in terms of metric units (e.g. the yeard is now defined as 0.9144 metre exactly). By 1969 the *Metrication Board* had been set up with the target of completing the metrication of British industry and commerce by 1975. This target was not met and the Metrication Board was disbanded in April 1980, metrication proceeding thereafter on a voluntary basis. In 1989 it was decided to keep Imperial measures for some items, including beer and milk, and to retain the mile, yard, foot, and troy ounce until at least 2000.

In science, former metric systems (such as the *c.g.s.* system and the *m.k.s.* system) have been replaced by *SI* units.

Metrology. The branch of science dealing the accurate measurement of the three fundamental quantities: mass, length and time. It is generally extended to mean the systematic study of weights and measurements.

Metronome. A small device consisting of a pendulum with a small sliding weight on it, which can be regulated to make the pendulum beat at a desired number of beats per minute. The most common type is the clockwork metronome invented by J.N. Maelzel (1770-1838); electric metronomes also exist. Metronome markings are often given in musical scores to indicate the exact speed of the music.

Mho. The reciparocal ohm, previously used as a unit of conductance. This unit is now replaced by the siemens.

Mica. A mineral made up of complex silicates, having a perfect basal cleavage enabling the crystals to be split into very thin plates. It has a low thermal conductivity and high dielectric strength, widely used for electrical insulation.

Michelson-Morley Experiment. An experiment performed by A.A. *Michelson* and E.W. *Morley* in 1881 in an attempt to demonstrate the existence of the luminiferous ether by measuring the earth's velocity relative to it. They used a Michelson interferometer to obtain interference fringes and then rotated the apparatus through 90^0 expecting to find a shift in the fringes, since the velocity of light would be different in the two directions. This difference would result from the earth's motion through the ether. However, no shift was detected. This negative result led to the downfall of the ether theory and was explained by Einstein's theory of *relativity in* 1905.

Microbalance. A balance that can weight very small masses (e.g. downs to 10^{-5} mg.). The beam is made to balance by changing the air pressure in the balance case so changing the upward buoyant force on a bulb fixed on one end of the beam. A manometer is provided for measuring the pressure when the beam is balanced; if the temperature is constant the density of the gass is proportional to the pressure.

Microcalorimeter. A differential calorimeter used in the measurment of very small quantities of heat, such as that evolved by a small quantity of radio-active substance.

Microdensitometer. An instrument for automatically measuring and recording small changes in transmission density across a sample, like a photographic plate.

Microelectronics. The branch of electronics dealing with the design, production, and application of electronic components, circuits, and devices of extremely small dimensions.

Micromanometer. An instrument for the measurement of very small pressure differences. The U-tube manometer has one arm of U nearly horizontal so that pressure changes causing only a small difference in vertical height produce easily visible movements of liquid in this arm.

Micrometer. An instrument for measuring small lengths with great accuracy. The object to be measured is held between the jaws of a C-shaped metal piece, one jaw of which can be adjusted by a screw. The screw is turned by rotating a drum with a *vernier* scale marked on it, from which the required dimension can be read.

Micrometer Eyepiece. An eyepiece, generally a Ramsden eyepiece, provided with cross-wires that can be displaced by means of a micrometer screw. It is used for the measurement of small objects.

Micrometer Screw. An instrument for measuring small and accurate distances. Such instruments are fitted with a drum which when rotated advances a screw of known pitch; the drum is calibrated in fractions of a revolution that can be interpreted in terms of the distance advanced by the screw.

Micron (μ). An obsolete name for one-millionth of a metre. The correct name is now the *micrometre.*

Microphone. A device that converts sound into electrical signals. It acts like a *loudspeaker* in reverse and some microphones may also be used as loudspeakers. Most consist of a thin diaphragm, the mechanical vibration of which is converted to an electrical signal proportional to the sound pressure. A telephone mouthpiece usually consists of a carbon microphone in which the sound waves exert a varying pressure on carbon granules, so varying their electrical resistance. In capacitor microphones, commonly used in music recording, the diaphragm forms one plate of a capacitor, across which the sound waves produce a fluctuating potential difference. Crystal microphones rely on the piezoelectric effect. Other types use magnetic induction or magnetostriction or other electromagnetic effects. Ribbon microphones have a highly directional response. They consist of a thin strip of aluminium alloy in a strong magnetic field. The ribbon vibrates in the sound waves, inducing in itself an electromotive force proportional to its velocity.

Microscope. An optical instrument used for producing a magnified image of a small object. There are several distinct types, the most common being the compound microscope, which contains an objective lens system and an eyepiece system. It was invented in 1609 by a Dutch spectacle maker Zacharias Janssen (1580-c. 1638) and his father, but Robert *Hooke* gave the first extensive description of its use in biology in his *Micrographia* (1665). In a compound microscope the objective produces a real magnified image, which is further magnified by the eyepiece. At low *magnifications* the system is illuminated by a source the light of which is reflected through the specimen by a mirror. At high magnifications special illuminating systems are needed. The magnification, which may be up to a thousand, is limited by the *resolving* power of the lenses; the smallest detail capable of resolution by an optical microscope is about 0.2 micrometre. This can be increased by using an oil-immersion lens. Still higher magnifications are obtained by using shorter wavelength radiations as in the *electron microscope.*

A *photomicrograph* is a photograph of the image obtained using a microscope. This enables a permannent record to be kept and also enables ultraviolet radiation to be used for illumination of the specimen.

Microwaves. Electromagnetic radiation with wavelengths between 1 and 300 millimetres, lying between infrared rays and radio waves in the electromagnetic spectrum. They are used in *radar* and *microwave heating*. This method is used in the rapid cooking of food as the radiation penetrates to the interior of the food. The microwave photon is the same order of magnitude as the vibrational energy of atoms and molecules and therefore heats the interior directly, rather than by conduction from the surface. It is also used in sterilization and in drying wood, etc. Microwaves are generated by such devices as *magnetrons* and *klystrons*.

Mie Scattering. Refers to scattering of light by spherical particles of diameters comparable with the wavelength.

Mil. One thousandth of an inch.

Miller Effect. In a single-valve amplifier, the apparent capacitance between grid and anode is multiplied, because of feedback, by (1-A) where A is the complex gain of the state. A is usually large and negative in sign, a positive voltage at the grid giving rise to a negative voltage at the anode. The effect also occurs between the collector and the base of at ransistor amplifier.

Miller's Law. In a crystal, the sine ratio of four faces in a zone is a commensurable number. (A zone consists of a set of planes whose intersecting edges are parallel.) If P_1 P_V, P_2 P_3 and P_4 are the four faces, and θ_{12} is the angle between P_1 and P_2 etc., then the sine ratio is given by

$$\frac{\dfrac{\sin\theta_{12}}{\sin\theta_{13}}}{\dfrac{\sin\theta_{42}}{\sin\theta_{43}}} = \frac{\dfrac{U_{12}}{U_{13}}}{\dfrac{U_{42}}{U_{43}}} = \frac{\dfrac{V_{12}}{V_{13}}}{\dfrac{V_{42}}{V_{43}}} = \frac{\dfrac{W_{12}}{W_{13}}}{\dfrac{W_{42}}{W_{43}}}$$

where $U_{12} = (k_1 l_2 - l_1 k_2)$; $V_{12} = (l_1 h_2 - h_1 l_2)$; $W_{12} = (h_1 k_2 - k_1 h_2)$ etc. *hkl* being the *Miller Indices* of the faces.

Milliameter. An instrument that can measumre electric currents of the order of a few millampers.

Minute. (1) A unit of time. 1 minute = 60 second.

(2) A unit of angle (symbol'). $1' = \frac{1^0}{60}$.

Mirage. An optical illusion mostly seen when the refractive index of different laysers of earth above its surface is not uniform. Light from object in the upper colder layers is refracted and finally undergoes total internal reflection from the warm, rarer layer close to the earth's surface and form an image e.g., the illusion of a tree nearly in a desert.

Mirrors. Devices for reflecting light, usually consisting of a sheet of glass with one surface silvered. A plane (flat) mirror forms a laterally inverted virtual image. Spherical mirrors, concave or convex, magnify or reduce the image. If the distances of the object and image from the mirror surface are u and v, then $1/u + 1/v = 1/f$, where f is the focal length of the mirror, taking all distances as positive in front of the morror and negative behind it. To avoid spherical aberration, parabolic mirrors are used in reflecting *telescopes.*

MIRV (Multiple Independently-targeted Re-entry Vehicle). A ballistic missile with several separate nuclear warheads, which can be aimed at different targets.

Mixed Grouping of Cells. An arrangement of cells where in combination of cells in series are connected in parallel. This arrangement yields maximum current if the external resistance in the circuit is equal to the total internal resistance.

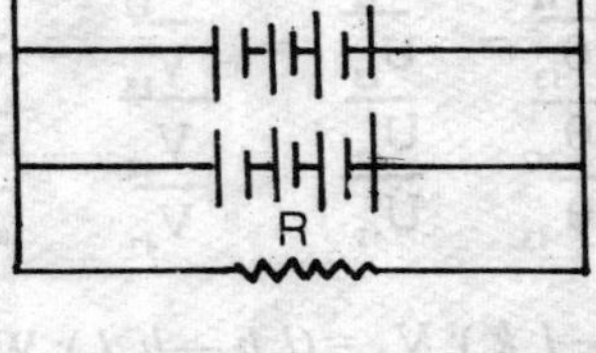

Mode. (1) One of several wave frequencies which an oscillator can generate or to which a resonator can be tuned. (2) The value of the abscissa corresponding to the maximum ordinate of a frequency distribution curve.

Modem. A device which converts the signals from one particular type of equipment into a form suitable for use in another.

Moderator. A substance used in nucleare reactor to slow down fast moving neutrons, which is essential to produce nuclear reactions. They are substances having low atomic weight such as carbon, paraffin wax, heavy water, etc.

Modulation. A method of carrying information (the signal) on an electromagnetic wave or an oscillating electric current. In *amplitude modulation* (AM) the amplitude of a carrier wave is changed according to the magnitude of the signal. This is used in medium-wave sound broadcasting in which audio-frequencies (50-20 000 hertz) are carried on radio waves with a frequency of about one megahertz. In *frequency modulation* (FM) the frequency of the carrier wave is changes within a small bandwidth of the reference frequency. FM is used in VHF *radio* (about 100 magahertz). It offers a better signal-to-noise ratio than AM.

In *pulse modulation* the carrier is a series of pulses. It is used in digital equipment, such as computers, and in telegraphy and telemetry. A continuous signal alters the height in pulse-amplitude modulation, the width in pulse-duration modulation, or the time between pulses in pulse-position modulation. Pulse-code modulation uses a coded pattern of pulses to carry the signal, e.g. in *Morse* code.

Modulus. (1) The absolute value of a number. The absolute value of — 8 is 8. (2) The magnitude of a vector quantity. The modulus of the velocity of a particle, is equal to its speed.

Mohs' Hardness Scale. A scale of hardness of materials based on a scratching test wherein any material can be scrateched by the material which follows it in the scale.

Molecular Cloud. An immense cloud of cool dense matter in the *interstellar* medium, composed mainly of molecular hydrogen with a very small proportion of other molecules and dust grains. A typical cloud is 130 light years across, has a mass of 500 000 solar masses, and a temperature of 10-20 K. Stars form in these regions and have been detected by radio astronomers in our Galaxy and others. Several thousand molecular clouds are thought to exist in the Galaxy.

Moment. The product of a force and its perpendicular distance from the axis about which it acts. A moment produces a turning effect and is sometimes called a torque. The inertia of a body to a torque is called its *moment of inertia.* This quantity is equal to mr^2 for a single mass (m) rotating about an axis at a distance r from the axis. The moment of inertia of a system of masses is equal to the sum of these products.

Moment of Inertia (I). The moment of inertia of body rotating about an axis is given by

$$I = \sum_{i}^{l} m_i r_i$$

where *mi's* are the masses of the particles in which the body can be divided and *ri's* their respective distances from the axis of rotation.

Momentum. The linear momentum of a body is the product of its mass and its linear velocity. The angular momentum of a body is the product of its *moment* of inertia and its angular velocity. Momentum is an important quantity in physics as during any process, for example a collision between two bodies, the total momentum of the system always remains constant (the law of conservation of momentum).

Monochromatic Radiation. Electromagnetic radiation with only one wave length or frequency. It is very difficult to produce purely monochromatic radiation and the term is often used to mean radiation having a very narrow range or wave lengths.

Monochromator. A device that gives monochromatic radiation when a radiation of many wave lenghs is fed into it.

Moon. The natural satellite of the earth. The moon orbits the earth every 27.32 days at a mean distance of 384 400 km, keeping more or less the sam face (the nearside) towards the earth. As it revolves, different *phases* can be seen from earth, together with up to two or three lunar *eclipses* per year. The moon is only 81 times less massive than the earth and has a diameter of 3476 km. Its origin is uncertain; of the various theories the most common are that it fomed from material that separated from the earth, from the debris created when a planet-sized mass collided with the earth, or from a separate body captured by the earth's gravitation.

The major surface features are the light-coloured highlands on the southern nearside and most of the farside, and the much darker lava plains—the maria. The maria and more especially the highlands are heavily cratered. These roughly circular walled depressions, ranging greatly in size, were produced by impacting bodies from space. The extremely tenuous atmosphere exposes the surface to considerable temperature extremes (—180°C to + 110°C).

Much of our information about the moon has been derived from photographs and other measurements taken from orbiting US and Soviet satellites and later from moonrock samples brought back (1969-72) by the Apollo astronauts (and the unmanned Soviet Luna landers) and from experiments set up on the moon by the astronauts. The first landing on the moon was made by Neil *Armstrong* and Edwin *Aldrin* on 20 July, 1969.

Morse Code. The code invented by Samuel Morse for transmitting telegraph messages. Each letter of the alphabet and number has a characteristic sequence of dots and dashes (short and long pulses), a dash being three times as long as a dot.

letters

A. —	H ....
B — . . .	I ..
C —. —.	J .— — —
D —. . .	K —. —
E .	L . — ..
F . . —.	M — —
G — —.	N-.

O — — —

P . — — .

Q — — . —

R . — .

S . . .

T —

U . . —

V . . . —

W . — —

X — . . —

Y — . — —

Z — — . .

numbers

1 . — — — —

2 — . . .

3 . . . — . —

4 . . . . —

5 . . . . .

6 — . . . .

7 — — . . .

8 — — — . .

9 — — — — .

0 — — — — —

punctuation marks

— . — . — .

, — — . . — —

: — — — . . .

? . . — . .

' . — — — — .

— — . . . . —

/ — . . — .

(or) — . — . —

" . — . . — .

Morse Code

Morse Equation. The potential energy of a diatomic molecule (Φ_r) according to P.M.Morse (1929) may be approximated by

$$\Phi_r = D\{1 - \exp[-a(r - r_e)]\}^2$$

where D is the dissociation energy, *r* is the distance apart of the atoms and r_e is the distance at equilibrium and *a* is a constant.

Moseley's Law. If X-rays are allowed to fall upon a substance, a series of characteristic x-ray lines is produced. Moseley showed that the frequency *v* of the line belonging to any particular X-ray series is quantitatively related to the atomic number Z by the relationship.

$$\sqrt{v} = a(Z - \sigma)$$

where *a* is the proportionality constant and σ is the same for all lines of a given series.

Mossbauer Effect. The Mossbauer effect is the recoil-less emission and resonant absorption of gamma radiation. When a nucleus decays from an excited state by emitting a gamma ray, the energy of the gamma ray is given by the energy level of the excited state less the energy imparted to the nucleus on recoil. The effect of the recoil of the nucleus is to reduce the gamma ray energy below the inter-level energy, such that a second nucleus in a lower state cannot be eexcited by gamma ray absorption. However, in 1958, Mossbauer found that, in certain systems (e.g. containing Fe^{57}), the recoil momentum is taken up by the whole crystal in which the nucleus is embedded. In this case, because of the much larger recoil mass, the energy of the emitted gamma ray is sufficient for it to excite a further nucleus. The uncertainty-spread in the frequency of the gamma ray is then as small as 1 part in 10^{13}. Hence the resonant absorption condition will be very sensitive to any relative velocity imparted between source and absorber, and a relative velocity of 1-10 mms^{-1} can throw the absorber out of resonance. Thus using separate source and absorber, the effect can be used for energy resolution when, for instance, the energy levels of the source are split by the *Zeeman Effect,* and can also be used to repeat the Michelson—Morley experiment.

Motorcycles. Two-wheeled engine-powered vehicles. The evolution of the motorcycle has been closely associated with that of the *bicycle, steam engine* and *car*. The concept of a steam-powered bicycle was first realized by S.H.Roper in the USA in the 1860s. Similar machines were being built in Paris at about the same time by Pierre and Ernest Michaux. However, the true forerunner of the modern motorcycle was Gottlieb *Daimler's* 1885 bicycle powered by an *Otto* four-stroke engine. The first production model was Hildebrand and Wolfmuller's 1894 *Petrolette*. By 1900 there were some 11 000 motorcycles in France and by the start of World War I over 100 000 were registered in the UK. By then the invidious belt drive (which slipped in wet weather) had been replaced by the chain, hot-tube ignition (which often set the whole machine alight) had given way to the sparking plug, and the battery had been augmented by magneto. During the war, motorcycles (often with sidecars) were extensivley used by both sides. The interwar period saw the development of many classic designs: the Horley-Davidson in the USA: the Brough Superior, Triumph Speed Twin, and Ariel Square Four in the UK; and the German DKW two-stoke and BMW four-stroke. All these were in military use in World War II.During and after the war innovations included telescopic forks, sprung near wheels, disc brakes, and starter motors. In the 1950s and 1960s interest in motorcycles in Europe and the USA declined; their inherent danger and the status symbol of car ownership forced most European manufacturers out of business. The motor scooter, a low-powered Italian version of the motorcycle, and the moped, an engine assisted bicycle, acquired some popularity. In the 1970s, the Japanese, exploiting the closure of European factories, developed a new range of motorcycles, based on European designs. These machines now dominate world markets.

Mott Scattering. Coulombic-type scattering by nuclei in which transverse polarization of the scattered electrons is obtained.

Mott Transition. A type of sharp transition from non-metallic to metallic behaviour. The transition is a consequence of the interaction of the electrons, as this interaction can give rise to insulating behaviour even if the electrons do not fill an energy band, provided the band is narrow. If the band is widened, for example by decreasing the interatomic spacing, transition to a conducting state may occur.

Moving Coil Instrument (Moving-coil Ammeter, Galvanometer, Voltmeter etc.) An electric instruments wherein rectangular coil placed between the pole pieces of a permanent magnet rotates when a current is passed through it. The coil wherein comes to rest when the rotating couple due to the magnetic field is equal to the restoring couple due to torsion in a suspending wire or a spring. The amount of deflection when the coil comes to rest is proportional to the amount of current flowing through the coil. A galvanometer, an ammeter and a voltmeter (d.c.) are all based on this principle.

Moving-iron Instrument. An electric measuring instrument in which an electric current is pased through a fixed coil, that attracts a soft iron piece, whose movement is dectected by a pointer attached to it. The movement is approximately proportional to I^2, (I being the current). Such instruments can be used for *a.c.* as well *d.c.*

Multimeter. An instrument that can be used to measure various electrical quantities like resistance, voltage (*a.c.* as well as *d.c.*), currents etc.

Multiplexer. In telecommunications, a device that combines several signals so that they can be sent along a single transmission path, or channel, and reconstructed at the receiver. One method is to superimpose carrier waves of different frequencies on the signals. Multiplexers are used with radio transmissions, telephone lines, etc.

Muon. A negativley charged unstable elementary particle (lifetime 2×10^{-6} second; mass 207 times that of the electron) that decays into an electron and two *neutrinos*. It has a corresponding antiparticle. It was originally thought to be a meson (and was called the mu-meson) but is now classified as a lepton.

Mutual Capacitance. The extent to which two capacitors can effect each other, expressed in terms of the ratio of the amount of charge transferred to one, to the corresponding potential difference of the other.

Mutual Inductance, Coefficient of (E). The *e.m.f.* induced in one circuit when the current in the second varies at the rate of 1 ampere/second. It is measured in Henry (H).

Mutual Induction. When the current or magnetic field associated with a circuit is varied an *e.m.f.* is produced in a neighbouring circuit. This is known as the phenomenon of mutual induction.

Myopia (Short Sightedness). Inability of the eye to clearly see distant objects. The image of a distant object is formed in front of the retina the defect can be a corrected by using a concave *lens of appropriate* negative power.

N

Nano (*n*). A prefix meaning 10^{-9}.

Near Point. The point nearest to the eye at which an object can be distinctly seen. Mostly the average near point distance is taken to be 25 cm.

Neel Temperature. The temperature at which the spins of the neighbouring ions in an antiferromagnetic material cease to be ordered antiparallel as the temperature of the material is raised.

Negative Crystals. Crystals in which the velocity of extra-ordinary ray is greater than the velocity of ordinary ray e.g. Calcite.

Negative Electron. *Syn.* negatron: negatron. An electron having a negative charge as opposed to the positively charged electron or positron.

Negative Resistance. The point at which the voltage-current characteristic has a negative slope, e.g. silicon-controlled rectifier, the magnetron, and the tunnel diode.

Neon Tube. A gas discharge tube filled with neon at low pressure, the colour of the glow being red.

Neper. Symbol N or Np. A logarithmic unit used for comparing two currents, generally those entering and leaving a transmission line or other transmission network. Two currents l_1 and l_2 are said to differ by N nepers when $N = \log_e [l_1/l_2]$.

Nernst Approximation Formula. Nernst was able to deduce the equlibrium constant of a gaseous reaction thermodynamically. From this relation, by a series of aproximations which are now known to be inaccurate, he was able to obtain an equation which, though crude, can give useful information when no other data are available. The equation is

$$\log K' = \frac{-\Delta H}{4.58\,T} - 1.75\,\Sigma V_1 \log T + \Sigma V_1\, C_1$$

where ΔH may be taken as the value of enthalpy measured at room temperature; Σv_1 is taken for the products minus reacting susbstances. For heterogeneous reactions, Σv_1 and $\Sigma v_1 C_1$ refer only to the gases, but ΔH to the whole reaction. C_1 is the conventional chemical constant, an empirical quantity and v_1 is the number of reacting molecules of species 1.

Nernst Effect (or Nernst-Ettingshausen Effect). If a temperature gradient is established in a conductor which is placed in a magnetic field that is orthogonal to the temperature gradient, then electrons tending to move fron the hot to the cold end are initially deflected by the magnetic field. This produces a transverse electric field which prevents further deflection and which is given by

$$E_y = QB_z = \frac{dT}{dx}$$

where E_y is the Nernst electric field, Q the Nernst coefficient, B_z the magnetic flux density and dT/dx the longitudinal temperature gradient.

The above effect is sometimes referred to as the transverse Nernst-Ettingshausen effect. The variation of the *Seebeck Effect* in a magnetic field is then called the longitudinal Nernst Ettingshausen effect.

Nerst Heat Theorem. Or the third law of thermodynamics. For a chemical change occurring between pure crystalline solids at absolute zero, there is no change in entropy.

Nernst-Simon Statement (of the Third Law of Thermodynamics). The entropy change associated with any isothermal reversible process of a condensed system approaches zero as the temperature approaches zero. It is sometimes called the Nernst heat theorem.

Nernst-Thomson Rule. J.J. thomson and W. Nernst in 1893 pointed out that, in a medium of low dielectric constant, the electrostatic attraction between the anions and cations of a dissolved electrolyte will be large, whilst in solvent of high dielectric constant the attraction will be small. Hence solvents of high dielectric constant will favour dissociation, whilst solvents of low dielectric constant will have small dissociating influence.

Nernst Zero of Potential. An arbitrary zero of electrode potential, defined as that potential corresponding to the reversible equlibrium between hydrogen gas at 1 standard atmosphere pressure and hydrogne ions at unit activity.

Neumann - Kopp Rule. The heat capacity of a gram-atom of a solid phase can be considered as the weighted sum of the heat capacities of the elements forming the phase. The rule is applicable to compounds formed from elements in the solid or liquid states.

Neumann's Law. Whenever the flux of magnetic induction enclosed by a circuit is changing, there is an e.m.f. acting round the circuit, in addition to the e.m.f. of any batteries which may be in the circuit. The amount of this additional e.m.f. is equal to the rate of diminution of the flux of induction enclosed by the circuit.

Neumann's Principle. The symmetry elements of any physical property of a crystal must include the symmetry elements of the point group of the crystal.

Neutral Filter. A light filter which absorbs equally all wavelengths, it reduces light intensity without any change of relative spectral distribution.

Neutral Point (Magnetism). A point in a magnetic region at which two magnetic fields are equal and opposite. The magnetic neutral points will exist, near a magnet placed in the earth's magnetic field.

Neutral Temperature. The temperature of the hot junction of a thermocouple at which the thermo e.m.f. around the circuit is maximum and the rate of change of e.m.f. with temperature is minimum.

Neutrinos. A group of three elementary particles and their antiparticles. They are classified as leptons, have no charge, and are probably massless. One type of neutrino is associated with the *electron*, one with the *muon*, and one with the *tau* particle.

Neutron. An elementary particle that is a consitituent of all atomic nuclei except hydrogen - 1. It has no electric charge and its mass is slightly greater than that of the *proton*. Inside the nucleus the neutron is stable but when free it decays by the *weak* interaction to a proton, an electron, and an antineutrino (*see* beta decay). Its mean life is about 12 minutes. The neutron was discovered by *Chadwick* in 1932.

Neutron Age. Refers to one sixth of the mean square displacement of a neutron as it slow down, through a specified energy range, in an infinite homogeneous medium.

Neutron Flux. Syn. neutron flux density. It is the product of the number of free neutrons per unit volume and their mean speed. The neutron flux in a power reactor lies of the range 10^{16}—10^{18} per square metre per second.

Neutron Number (N). Refers to the number of neutrons present in the nucleus of an atom.

Newton (N) The *SI* unit of force defined as the force required to give a mass of one kilogram an acceleration of one metre per second per second. named after Sir Isaac *Newton.

Newtonian Fluid. A fluid wherein the amount of strain is proportional both to the stress and to the time. The constant of proportionality is known as the coefficient of viscosity.

Newtonian Force. Syn. Coulomb force. A force between points that falls of as the inverse square of distance between them.

Newtonian Mechanics. The branch of *mechanics* concerned with systems in which the results of *quantum* theory and the theory of *relativity* can be ignored. Also known as classical mechanics, it is based on *Newton's* three laws of motion. The first law states that a body remains at rest or moves with constant velocity in a straight line unless acted upon by a *force*. This law thus defines the conept of force. The second law, which defines mass, states that the *acceleration* (a) of a body is proportional to the force (f) causing it. The constant of proportionality is the mass (m) of the body: $f = ma$. The third law states that the action of a force always produces a reaction in the body. The reaction is of equal magnitude but opposite in direction to the action.

Newtonian and Non-Newtonian Liquids. Liquids which show a linear relationship between velocity gradient and shearing stress are termed Newtonian liquids. Complex substances and high molecular-weight and colloidal solutions do not exhibit this simple behaviour; the velocity gradient increases superlinearly with shearing stress, i.e. the viscosity of the liquid decreases with shearing stress. Such liquids are called non-Newtonian.

Newton's Equation (for Conjugate Distances). For a perfect lens system,, if f and f' are the first and second focal lengths of the system, corresponding to space to the left and right of the system respectively, and x and x' are the distances of the objects and image from the foci F and F' respectively, as measured outwards form the lens system, then algebraically

$$xx' = ff'$$

Newton's Equation (for the Velocity of Sound). This equation may be expressed in the form

$$v = \frac{p}{\rho}$$

where v is the velocity of sound in a gas at pressure p and of density ρ and is inaccurate as it assumes isothermal conditions.

Newton's Law of Cooling. The rate of loss of heat from a hot body is directly proportional to the difference of temperature over that of its surroundings, if the excess of temperature is not very large.

Newton's Law of Gravitation. Everybody in this universe attracts every other body with a force that is directly proportional to the square of the distance between them. Mathematically

$$F \alpha \frac{m_1 m_2}{r_1} \text{ or } F = G \frac{m_1 m_2}{r_2}$$

where m_1, m_2 are the masses of two bodies and r is the distance between them. G is a constant, known as the universal gravitational constant and has a value 6.67×10^{-11} N m^2/kg^2.

Newton' Law of Resistance. At moderate velocities the resistance of a medium is proportional to the square of the velocity.

Newton's Laws of Motion. (1) Every body continues in its state of rest or of uniform motion in a straight line except in so far as it may be compelled by impressed force to change that state. (2) Change of motion is proportional to the impressed force and takes palce in the direction of the straight line in which the force acts. (3) To every action there is an equal and opposite reaction.

Newton's Rings. A series of light and dark rings formed in a plano-convex lens if monochromatic light is shone onto the lens when it rests on a plane mirror. First observed by *Newton.*, they are caused by *interference* between light reflected by the mirror and light reflected at the curved surface of the lens.

Newton's Theory of Light. Newton enunciated the corpuscular theory of light in 1678. A flight of material particles was emitted by a light source and the impact of these particles on the retina produced the sensation of sight. Rectilinear propagation is a consequence of Newton's second law. Refraction and reflection were explained as due to forces of repulsion and attraction exerted, at the surface of the reflecting or refraction medium, on the particles.

Nickel-iron Accumulator. (Ni-Fe cell): A secondary cell in which the positive plate is nickel-oxide and the negative plane is of iron. The electrolyte is a solution (20%) of potassium hydroxide. The cell e.m.f. is nearly 1.3 volts. It is more durable and lighter as compared to the lead accumulator. *Also known* as Edison Cell.

Nicol Prism. A prism made of calcite used for polarizing light and analysing plan polarised light.

Nit. Symbol: nt. A unit luminance, defined as one candela per metre squared. The unit is identical with the SI unit of luminance.

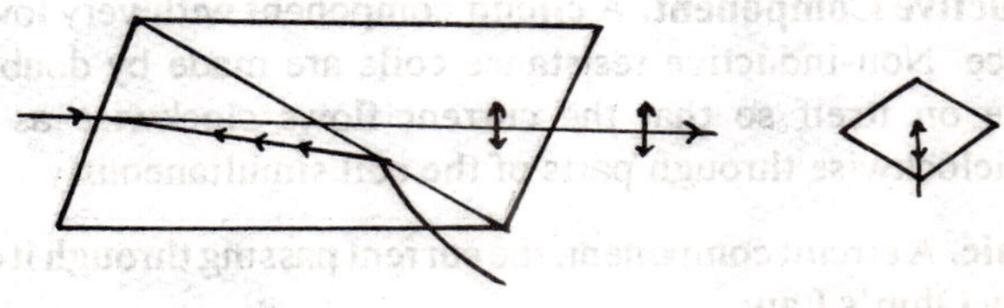

Nodal Line. Line that joins the positions of minimum disturbance (nodes) in an interference pattern.

Nodal Points. A pir of points on the principal axis for which there is unit positive angular magnification. An incident ray directed towards one nodal point emerges parallel to itself through the other nodal point.

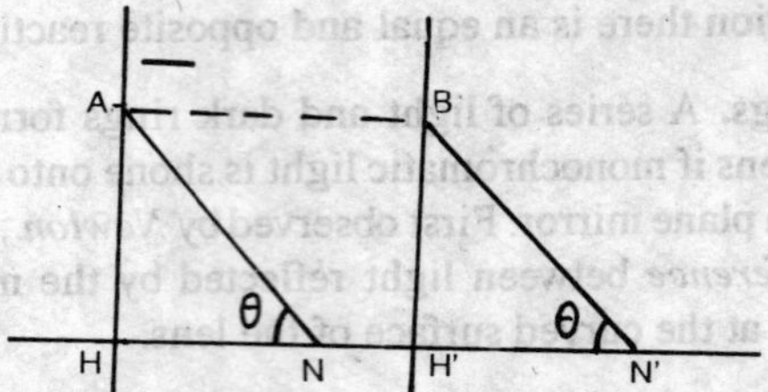

In the figure N, N, are the nodal points. H.H' are the principal points whereas planes AH, BH, are principal planes.

Node. A point in a stationary wave pattern having the minimum vibration.

Noise Factor. A measure of the noise. It is defined as the ratio of the signal-to-nose power ratio at the source to the corresponding ratio at the output.

No-load. Refers to operation with no output but under rated conditions of voltage, speed etc.

Nondegeneracy. Refers to the normal state of matter, i.e. matter not raised to a very high temperature nor subject to excessive stress such that the density is abonormally high. A nondegnerate gas is characterized by Maxwell's law of the distribution of velocities.

Non-inductive Component. A circuit component with very low inductance. Non-inductive resistance coils are made by doubling the wire on itself so that the current flows clockwise as well as anticlockwise through parts of the coil simultaneously.

Non-ohmic. A circuit component, the current passing through it does not obey Ohm's Law.

Nordheim's Rules (for Nuclear shell Model). Rules for the shell model of the nucleus applicable to nuclei which have odd numbers of protons and neutrons. The Nordheim number is defined as

$$N = j_p - l_p + j_n - l_n$$

where l is the orbital angular momentum and j is the total angular momentum. Subscripts p and n refer to the odd proton and neutron respectively. Then the rules state;

$$\text{if} \quad N = 0, I = | j_n - j_p |$$

$$\text{If } N = \pm 1, I = | j_n - j_p | \text{ or } | j_n + j_p |$$

where i is the total angular momentum of the nucleus. There are exceptions to the rules, especially among the light nuclei.

N.T.P. Normal temperature and pressure. Their values are respectively 27° K and 76 cm. of mercurry.

n-type Semiconductor. An extrinsic semiconductor whose density of conduction of electrons exceeds that of mobile holes.

Nuclear Barrier. Syn. Gamowbarrier. A region having potential energy so that a charged particles must pass through in order to enter or leave an atomic nucleus.

Nuclear Energy. The energy given out during a nuclear fusion or fission process.

Nuclear Fission. A nuclear reaction wherin an atomic nucleus disintegrates into parts releasing a large amount or energy. The process may be pontaneous or may be due to the impact of a particle.

Nuclear. A strong attractive force that exists between the nucleons. It is short-range force having a range or a few fermi (fermi = 10^{15} m) and is charge-independent.

Nuclear Fusion. A nuclear reaction wherein two light nuclei fuse to from a heavy nucleus along with the release of large amount of energy.

Nuclear-physics. Branch of physics that deals with the structure and properties of nuclei, nuclear reactions and nucleon-collisions.

Nuclear Reaction. A reaction wherin there is a change in the nucleus of an atom; it involves nuclei, nucleons, gamma rays and sometimes lighter elementary particles.

Nuclear Reactor. A device wherin the energy produced in a nuclear fission reaction is used as useful energy. It consists of a core contaînning the nuclear fuel (generally uranium) that undergoes fission producing two nuclides or approximately equal mass with the release of 2-3 neutorns and as large amount of energy. These neutrons cause fission in the remaining fuel and a chain reaction takes place. *Control rods* in the reactor absorb some of these neutrons and help in maintaining the reaction at a constant level. A *moderator* is used to slow down the fast moving neutrons, a process which is essential for producing more fission reactions and thus sustaining the chain reaction.

Nucleor. The postulated core of a nucleon, supposed to be surrounded by a cloud poins.

Nucleus, Atomic. A very small region of space at the centre of an atom containing the nucleons in atom around which the electrons move in closed orbits. The radii or different nuclei are in the range 10^{-15} —10^{-14} m. The density of nuclear matter is very high (nearly 10^{15} kg/m^3). A nucleus is mostly written Xz^A where X is its chemical name, Z is the number of protons, it contains and A is its atomic mass number i.e., the number of protons and neutrons in the nucleus.

Nuclide. Refers to a nucleus of an atom, having a given number of nucleons. Thus O^{16} and O^{17} are different nuclides.

Nyquist Rule. If a response-vector locus be plotted for a servo system for positive and negative frequenscies and if the point (— 1, j0) is not enclosed, the system is unstable.

Nyquist's Theorem. The mean-square-voltage across a resistance R in thermal equlibrium at absolute temperature T is given by

$$\overline{V}^2 = 4R\, k_B\, T\, \Delta f$$

where Δf is the frequency band within which the voltage fluctuations are measured.

O

O.A.S.M. System. A system of units based on the ohm, ampere, second and metre.

Object. Natural point objects self-luminous or otherwise, deliver diverging rays of light.

Objective. A lens or a system of lenses near the object in microscope or a telescope. Also called the object lens.

Observatory. A collection of optical, *infrared,* or *radio* telescopes together with anciliary equipment and housing. Observatiories for optical and infrared studies are now sited at high altitudes so that atmospheric disturbance is minimized. Space observatories, orbiting above the earth's atmosphere, permit X-ray, gamma-ray, ultraviolet, and infrared studies to be made and improve optical measurements.

Occultation. The cutting off of light from a celestial body when another body intervenes the observer and it.

Oetave. The gap between two musical notes whose frequencies are in the ratio 2 : 1.

Octet. Group of eight electrons that constitute the outer electron shell of an inert gas (except helium) or an atom.

Odd-Even Nucleus. Nucleus having an odd number of protons and even number of nucleous.

Odd-Odd Nucleus. Nucleus having odd number of protons as well as of neutrons.

Oersted. C.G.S. unit of magnetic field strength. Its value is $\frac{10^3}{4\pi}$ A/ m.

Ohm. (Ω). The *SI* unit of electrical resistance equal to the resistance between two points on a conductor when a potenital difference of one volt between the points produces a current of one ampere. This definition replaced the former definition, which was based on the resistance of a specified column of mercury. Named after Georg *Ohm.*

Ohmic Contact. A contact in which the potential difference across it is directly proportional to the current flowing throught it.

Ohm's Law. The basic law of electric current, named after its discoverer Georg *Ohm*. The current, I, flowing through an element in a circut is directly proportional to the voltage drop, V, across it. It is written as V = IR, where R is the resistance of the circuit element.

Ohmic Loss. Refers to power dissipation in an electrical circuit arising from its resistance rather than from other causes like back e.m.f.

Ohmic Substance. Refers to a substance, the current flowing through which is in accordance with Ohm's law.

Opacity. Reciprocal of the transmittance of a substance. It is measure of the extent of opaqueness of a substance.

Opaque. Any susbstance that does not allow light to pass through it is called an opaque susbstance.

Open-circuit. Refers to an electric circuit through which no current is passing.

Optical Activity. The property of rotating the plane vibration of polarized light possessed by certain substances and their solutions. The rotation produced depends on the length of the medium and the concentration of the solution, and also the wave length of light used.

Optic Axis. A direction in a doubly refracting crystal in which light propagates without under going double refraction.

Optical Axis. It is a line passing through the optical centre and centre of curvature of a spherical lens or mirror.

Optical Centre. Refers to the geometrical centre of lens through which an incident ray passes undeviated.

Optical Flat. A surface that is so flat that irregularities of surface flatness are not greater than a fraction of the wavelength of light. The surface can be tested by obsorving interference fringes when a known flat is placed in contact.

Optical Glass. Glass having no mechanical and optical defects (Density, strain, heterogreneity, colour, refractiv index, etc).

Optical Lever. A device to measure regular displacement of an object. It consists of a small mirror fixed on a body capable of rotation. A narrow beam of light is allowed to fall on it and after reflection from the mirror on a scale. If the body suffers an angular displacement, θ the reflected ray turns through 2θ and the reflected light falls at a different point on the scale. The angle θ can be measured from the difference in the readings on the scale.

Optical Path. Refers to the product of the refractive index of the medium and the distance traversed by the light in the medium.

Optical Pyrometer. A pyrometer wherein the luminous radiation from the hot body is compared with that from a known source. The instrument measures the temeprature of a lminous source without thermal contact.

Optical Pyrometer. *See* pyrometer.

Optics. The branch of physics concerned with *light* and *vision*. Optics is divided into two major branches: geometrical optics and physical optics. Geometrical optics studies the geometry of light rays as they pass through an optical system.

Physical optics concerns light's properties (e.g. *diffraction, interference* and *polarization*) and its interaction with matter (e.g. in *refraction, scattering* and *absorption*).

Orbital Velocity. Refers to the velocity required by a satellite or spacecraft to enter and maintain a particular orbit around the earth or some other celestial body. The orbital velocity needed for a 24-hour orbit around the earth is approximatley 3.2 kilometres per second, at an altitude of about 36000 km.

Ordinary Ray. For double refracting crystals if the angle of incidence of a ray is changed, then only one of the two refracted rays obey Snell's law. This is called the *ordinary ray*. For the other ray, the ratio $\frac{\sin i}{\sin r}$ = n is not constant. The second ray is called *extra-*

ordinary ray.

Organ Pipe. An air column (enclosed in a pipe) having one or both ends open. It is called an open pipe if both ends are open and a closed pipe if one end is closed . The ends of an open pipe are always antinodes and the closed and of a closed pipe is a node.

Orthicon. A television camera tube in which a low-energy electron beam scans a target screen, consisting of a thin dielectric plate with photosensitive squares on one side and a thin metallic coating on the other. Each square and its metallic electrode form a tiny capacitor; when light falls on a photosensitive element the capacitor becomes charged. The scanning beam discharges these capacitors and thus becomes modulated by the pattern of light falling on the screen.

Orthochromatic Film. Photographic film which is sensitive to green as well as to the blue end of the spectrum.

Oscillator. A device that produces a.c. signals of a given frequency from a d.c. input. It usually consists of thermionic valves or transistors coupled with a suitable L-C-L resonance circuit.

Oscillatory Circuit. A circuit made up of a capacitor and an inductance. The frequency of electrical oscillations produced by such circuit is given by

$$f = \frac{1}{2\pi\sqrt{LC}}$$

where L is the inductance and C is the capacitance in the circuit.

Oscillation Magneto Meter. A device in which a freely suspended magnet is allowed to vibrate in a magnetic field (of earth). The time-period of vibration is given by the formula.

$$t = 2\pi \frac{K}{MH^2}$$

where H is the magnetic field, M is the magnetic moment of the magnet and K is a constant.

Oscillogram. Refers to the record produced by a recording oscillograph, or the reading from an oscilloscope.

Oscillograph. An oscilloscope particularly one equipped to make a permanent record of the parameter being measured.

Oscilloscope. A device used to produce a visual image of one or more rapidly changing electrical quantities. The cathode-ray oscilloscope is the most common type.

Ostwald Viscometer. An instrument for measuring or comparing the viscosities of liquids, made up of two bulbs at different heights connected by a capillary tube. The times t_1 and t_2 taken for each liquid to flow out of the upper bulb is noted. The viscosity is then given by $n_1 / n_2 = P_1 t_2 / P_2 t_1$, where p is the density.

Output. (1) It is the power, voltage or current delivered by any circuit, device or plant. (2) The terminals or other place where the signal is delivered.

Output Impedance. The impedance offered to the load by a circuit or device.

Output Transformer. Refers to transformer that is used to couple an output circuit to a load.

Overcurrent Release. *Syn.* overload release. A tripping device which operates when the current exceeds a predetermined value.

Overtones. The tones or frequencies emitted by a system in addition to its fundamental frequency are known as overtones. The intensity of overtones, is generally lower that of the fundamental.

Oxygen Point. Refers to the temperature of equilibrium between liquid and gaseous oxygen at a pressure of one standard atmosphere, taken as a fixed point the international temperature scale. at 90.188 kelvin.

P

Pair Production. Simultaneous production of an electon and its antiparticle (positron) from a gamma ray photon. The minimum energy such a photon must have is 1.022 Me V, which is equal to the sum of the rest masses of the electron and positron.

Parachuting. The use of a parachute (a fabric canopy) to float down to the ground, usually from an aircraft, either if the aircraft is about to crash, as a way of landing troops in an area, or as a form of sport. Parachutes were used for entertainment long before they were used for safety; the first successful jump was made in 1797 and the first contest held in 1926. In competitions contstants use vents in the parachute to steer onto a target area, a red cross with arms 5 m (16 ft) long and 1 m (3 ft) wide. In the middle is a red disc 10 cm (4 in) in diameter. Top-class parachutists land exactly on the disc.

Parallax. The apparent displacement of an object against a distant backgroud when viewed from different positions. In astronomy *diurnal parallax* is the angular displacement in the apparent position of a celestial body as a result of viewing it from the earth's surface rather than its centre. *Annual parallax* is caused by observations being made from the earth rather than the centre of the sun.

Parallel Resistance in (Capacitors in). Electrical resistances or capacitors are said to be connected in parallel when they are cannected in such a way that the current divides between them at one end and recombines at the other end. For resistances R_1, R_2,R_3 ...... joined in parallel, equivalent reistance R is given by

$$\frac{1}{R} = \frac{1}{R_1} + \frac{1}{R_2} + \frac{1}{R_3} + ...$$

The equivalent capacitance C of capacitannces C_1 C_2.......joined in parallel $C = C_1 + C_2 + ...$

Parallelogram Law of Vectors. The resultant of two vectors intersecting at a point is respresented by the diagonal of the parallelogram formed with the vectors as the two sides, and passing through the point of intersection of the vectors.

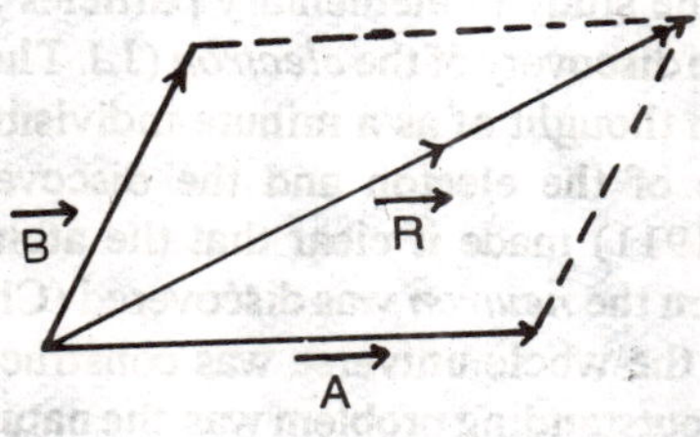

Paramagnet (Paramagnetic Substance). Susbstances that get weakly magnetised on being placed in a magentic field. When placed in non-uniform magnetic field, they move from weaker parts to the stronger parts of the field.

Parametric Amplifier. A low noise microwave amplifier wherein gain is otained by periodically changing a circuit parameter, e.g. capacitance.

Parasitic Capture. Capture of a neutron by an atomic nucleus without any consequent nuclear fission taking place.

Parasitic Oscillations. Unwanted oscillations that may take place in the circuit of an amplifier or oscillator. Such oscillations generally have a frequency that is very much higher than the frequencies for which the circuit has been designed since the frequency is mainly determined by the stray inductances and capacitances of connecting leads and by interelectrode capacitances.

Paraxial Rays. Those incident rays which are parallel and close to the axis of a lens (surface).

Parent Nucleus. A nucleous which undergoes radioactive decay to another nucleus, the resulting nucleus is known as the *daughter* nucleus.

Parsec. A unit of distance, used in astronomy, corresponding to a parallax of one second of arc. 1 parsec= 3.26 lightyears or 3.084

$\times 10^{16}$ metres.

Partial Pressure. The partial pressure of a gas in mixture of gases occupying a fixed volume is the pressure that the gas would exert if it alone is present in the total volume.

Particle Physics. The study of elementary particles and their interactions. Until the discovery of the *electron* (J.J. Thomson; 1897), the atom had been thought of as a minute indivisible "billiard ball." The existence of the electon and the discovery of the *proton* (Rutherford; 1911) made it clear that the atom had an internal structure. When the *neutron* was discovered (Chadwick; 1932), it appeared that the whole universe was constructed of these three particles. The outstanding problem was the nature of the force that held neutrons and protons together in the the nucleus. The only fundamental forces known at that time were the gravitational force and the electromagnetic (em) force: the gravitational force was too weak to account for the stability of the nucleus and the em force had no effect on neutrons.

In 1935 *Yukawa* suggested that there might might be in nature a short-lived particle (later called the meson) that jumped between protons and neutrons and held them together. This concept of exchange forces and the subsequent discovery of short-lived particles led to intensive research into particle physics (*see* accelerators). By the 1960s some 200 "elementary" particles had been identified and it became clear that there were four basic types of forces; in addition to gravitational and em forces there were *strong interactions* (10_0 times more powerful than em forces) and *weak interactions* (10^{10} weaker than em forces). In general, there are now belived to be two classes of elementary particles: leptons (the electron, muon, tau particle, and *neutrons*), which interact by the em or the weak forces and have no apparent internal structure; and and hadrons (including the proton, pion, etc.), which interact by the strong interaction and have an internal structure.

The current model of hadrons is based on Murray Gell-Mann's concept of the quark, introduced in 1964. In this model, hadrons themselves are divided into two classes: baryons, which decay into protons; and measons, which decay into leptons and *photons* or into proton pairs. Baryons consist of three quarks and mesons consist of a quark-antiquark pair. Thus all the matter in the universe is now seen as being made of leptons and quarks.

Although a single quark has never been identified experimentally, there is now some evidence that they actually exist. Quarks have fractional electronic charges (+2/3 or -1/3 of the electronic charge) and come in six "flavours" called up (u; +2/3), down (d; -1/3), charmed (c; + 2/3), strange (s; — 1/3), top (t; + 2/3), and bottom (b; - 1/3). For each flavour there is an equivalent antiquark ($\bar{u}$, $\bar{d}$, etc.). The proton consists of uud (2/3 + 2/3 — 1/3 = 1) and the neutron consists of udd (2/3 — 1/3 — 1/3 = 0).

In this form quark theory conflicted with the *Pauli* exclusion principle and it therefore became necessary to introduce colour charge. Thus each flavour of quark can have one of the three colours red, green or blue, with antiquarks having the corresponding anticolours. Colour charge has no connection with visual colour but the analogy is useful. All hadrons are regarded as white and baryons must consist of a red, a blue, and a green (since these visual colours produce white); measons consist of a quark of any colour and its correspondig anticolour. This aspect of particle physics is known as *quantum chromodynamics* by analogy with quantum electrodynamics. In this theory the strong interaction is thought to occur by the exchange of gauge bosons called gluons, just as them interaction occurs but the exchange of photons. Gluons are massless and chargeless by they have a colour charge. There are eight types, each carrying one colour and an anticolour. In a strong interaction a quark can change its colour but all colour changes involve the emission of a gloun, which is absorbed by another quark and causes its colour to change. Thus quark colours move from point to point, the strong interaction being seen as the force that maintains hadrons white.

Particle Velocity. Symbol: μ The alternating component of the velocity of a medium which is transmitting sound. It is the total velocity of the medium minus the velocity which is not due to sound propagation. As the velocity is changing regularly with time the root-mean-square value is generally taken.

Pascal's Principle. Pressure applied at any point of a fluid at rest is transmitted without loss to all other parts of the fluid.

Paschen-Back Effect. Paschen and back discovered in 1912 that all the simplicity of many Zeeman patterns was lost when the magnetic separation was comparable to the natural separation of a doublet, triplet, etc. The lines apeared to influence each other until, when the magnetic separation was large compared to the natural separation, the narmal *Zeeman Effect* returned both in pattern and in magnitude.

A similar effect, in which the hyperfine structure in a spectrum breaks up in a magnetic field but begins to overlap again as the field is further increased, is called the Back-Goudsmit Effect.

Paschen's Law. At a constant temperature, the breakdown voltage of a gas is a function only of the product of the gas pressure and the distance between parallel plane electrodes.

Peak Value. Refers to the maximum value of an alternating physical quantity.

Peclet's Number. A dimensionless number for a fluid flowing past a body and equal to *Prandtl's Number* × *Reynolds Number*.

Peierls Stress. The minimum energy required to remove a dislocation form its rest position of mimimum energy.

Peltier Effect. Heat is, in general, absorbed or liberated when an electric current crosses the junction between two metals A and B. The heat is proportional to the current i and is equal to Πi where Π is called the Peltier coefficient of A with respect to B and varies with temperature.

Pendulum. A device in which a mass (the bob) swings freely about a fixed point with a constant period. In the ideal simple pendulum the bob is connected to the fixed point by a length (l) of weightless string, wire, etc. Its period is $2\pi(l/g)^{1/2}$, where g is the *acceleration* of free fall, and is independent of the mass of the bob. A compound pendulum consists of a bob attached to the fixed point via two rigid rods. Pendulums are used to regulate a clock mechanism and in instruments that determine the value of g.

Penning Effect. The ionization of one species of gas molecule by collision with another species in a mixture of gases, when the former species is in a metastable excitation level which is above the ionization potential of the other.

Penrose's Theorem. A collapsing object whose radius is less than its gravitational radius must collapse into a singularity. The gravitational radius of an object is the radius which that object would have, such that light emitted from its surface just ceases to escape because of the gravitational field at that distance.

Pentode. A thermionic valve having five electrodes-cathode, three grids and an anode. The addition of a suppressor grid to the tetrode, kept at the cathode potential, helps to suppress the loss of secondary electrons given out by the anode.

Penumbra. Region of partial shadow, obtained when an object in the path of light from a source cuts off a part of light.

Perfect Gas. An ideal gas that would obey the gas laws exactly. It consist of perfectly elastic molecules, their volume and the intermolecular forces between them being zero.

Pericynthion. Refers to the time or the point at which satellite launched from earth into lunar orbit is nearest to the surface of the moon.

Perigee. Refers to the shortest distance of a stellite from the earth.

Peribelion. When the earth is nearest to the sun (Jan. 3rd) it is said to be in peribelion with respect to the sun. Conversely, the sun is in *perigee* with respect to the earth. When the earth-sun distance is maximum (July 4th), the corresponding terms are *aphelion* and *apogee*.

Perilune. The time or the point at which a satellite launched from the moon into lunar orbit and closet to the surface of moon.

Period (T). The time taken for one complete cycle of a periodic motion. It is equal to the reciprocal of the frequency of the motion.

Periodic Motion. Motion in which the body returns to the same condition of motion after regular intervals of time. e.g. Simple harmonic motion.

Perscope. An optical device consisting, typically, of a tube in which mirrors or prisms are arranged so that light passing through an aperture at right angles to the tube is reflected through the length of the tube to emerge at an aperture at the other end also at right angles to the tube . Periscopes in their simplest form are used to see over the heads of a crowd; in their more sophiscated form, they are extendible and used by submerged submarines to see above the surface of the water. In this form they may contain aiming devices for weapons, etc.

Permanent Magnet. A substance that possesses magnetic properties without the influence of any external magnetic field.

Permeability (μ). The ratio of magnetic flux density, B, in the susbstance to the applied magnetic field H. Thus $\mu = \frac{B}{H}$. It is measured in

henry/metre and is known as absolute permeability.

The relative permeability, is equal to the ratio of its absolute permeability to the permeability of the free space.Thus $\mu_r = \frac{\mu}{\mu_0}$

where, μ_0 the permeability of free space has the value $4\pi \times 10^{-7}$ henry/metre.

Permitivity. In accordance with Coulomb's law, the mutual force between two charges q_1 and q_2 at a distance r from each other is given by $F = \frac{1}{4\pi\epsilon} \frac{q_1 q_2}{r^4}$ where is a constant called the permeability of the medium. The permitivity of free space, ε_0 has the value 8.854 $\times 10^{12}$ farad/metre.

Perpetual Motion. Motion of a machine, which once set in motion goes on moving without receiving energy from any external source.

Persistence of Vision. The impression of an image on the retina of the eye for sometime after its withdrawal is known as persistance of vision. The impression on human eye lasts for $\frac{1}{16}$ of a second. Successive images at the rate of 16 per second of the same scene will give the impression of continuity. The principle is practised in cinema.

Petrol Engine. An Internal combustion engine wherein a mixture of vapour from petrol and air is oxploded in a cylinder with the help of an electric spark. Its complete cycle of operations is divided into four stages and it is called a four-stroke engine. Its efficiency is nearly 25%.

Phase. A stage in a cycle which a repetitive motion or wave has reached at a given time. Two waves are said to be in phase which their corresponding points coincide.

Phase Delay. Refers to the radio of the phase shift of a periodic quantity to its frequency.

Phase Diagram. A graph combining two conditional parameters (e.g., temperature, pressure, entropy, volume) or a substance drawn so that a particular curve shows the boundary between two phases of the substance.

Phase Difference. Refers to (1) The difference in phase of a wave at two points. (2) The difference in phase between two waves arriving simultaneously at a point in space.

Phase Shift. Any change that takes place in the phase of one quantity or in the phase difference between two or more quantities.

Phase Space. A multidimensional space wherein the coordinates represent the variables required to specify the state of the system, especially a six-dimensional space incorporating three dimensions of position and three of momentum.

Phase Splitter. A circuit wherein a single input waveform produces two output waves having a given phase difference, e.g., the driving circuit for a push-poll amplifier.

Phase Velocity. Refers to the velocity of propagation of phase in a travelling wave. Its value is equal to $\frac{\lambda}{T}$ where μ is the wave length of the wave and T is the time period of the wave.

Phasor. A rotating vector that is used for representing an alternating current or some other sinusoidally changing quantity. The projection of the phasor on a fixed line gives its amplitude at any time.

Phanon. The phanon is a quantum of thermal energy. It is given by hf, where h is the Planck constant and f the vibrational frequency. It refers to lattice vibration of crystals.

Phon. A unit for measuring the loudness of sound equal to the intensity in *decibels* of a sound of frequency 1000 hertz, which appears to the ear to be as loud as the sound to be measured.

Photocathode. A substance that emits electrons when light is incident on it, on account of photoelectric effect, and which acts as a cathode.

Photocell (Photoelectric cell). A device used for detection and mesurement of light. Its working depends on (i) photoelectric effect (ii) photovoltaic effect or (iii) photoconductive effect. The cell is known as a photoemissive cell, a rectifier cell or a conductivity cell respectively. The current flowing through the cell in the first case, the voltage developed across the terminals in the second case and the resistance of a layer of material in the third case, when the cell is exposed to light depends on the light incident on the cell. Their measurement enables one to calculate the light falling on the cell.

Photo Condutive Effect. The increase in the electrical conductivity of some substance when the amount of light falling on the susbstance is increased e.g. selenium.

Photocopying Machine. A device that prints black-and-white or coloured copies of document, drawings, etc., from an optical image. The most common technique is a dry electrostatic process known as xerography. A pattern of electric charge is induced by light falling on a layer of *semiconductor* material on a conducting surface. Toner powder is sprayed or rolled onto this material, so that it sticks to the hightly charged areas. The image so formed is printed onto charged paper, where it is fixed by heating to produce the final copy. Xerography is also used for high-speed printing of computer output, with printing heads instead of a light image.

Photo Disintegration. The disintegration of nucleus when a photon is incident on it. The nucleus generally emits neutrons or small charged fragments.

Photoelectric Effects. A number of effects in which electromagnetic radiation interacts with matter, frequently with the emission of electrons. These effects include *photoconductivity,* the *photovoltaic* effect, and the *Compton* effect and the *Auger effect.* The frequency of the radiation (f) has to be such that the energy of the photon (hf, where h is Planck's constant) is sufficient to liberate the electron. For solids, the mimimum energy requird is called the work function; for free atoms or molecules it is equal to the first *ionization* potential, the effect then being known as *photoionization*. For most substances, an ultraviolet frequency is needed to eject an electron, but for some metals, such as caesium visible light is sufficient.

Photography. The recording of images on sensitized material, by means of visible light, X-rays, or other radiation, and the susbsequent chemical processing. The first photograph was taken in 1826 by Joseph Nicephore Niepce (d. 1833) using the action of light on asphalt solution. In 1839, **Daguerre* introduced the daguerrotype, a positive image of milky white on a silver background, produced directly from silver iodide emulsion on plates exposed inside a simple *camera*. Modern photography involves a negative made by the development of *film coated* with silver salts. The positive picture is obtained by shining light through the negative onto light-sensitive paper with a coating similar to that on the original film. Lenses can be used to enlarge or reduce the size of the final image on the print. Transparencies, for use with a slide *projector*, may also be made. Film is usually developed and printed in a darkroom by dipping it in baths of chemicals. Polaroid photography, however, produces positive picutes directly from the camera in one stage. This system was invented by Edwin Land in 1947 and is used for instant picures for security passes, laboratory records, etc.

Photography as an art dates back almost to the origins of the technique. Portaiture was already established in the 1850s and was a prime interest of hte Photographic Society (later the Royal Photographic Society) founded in London in 1853. The Crimean War in the 1850s and US Civil War in the 1860s provided the first opportunities for the true miseries of war to be realistically portrayed. However, during the last half of the 19th century photographic interest centered on the landscape, especially as horizons broadened with the growth of travel.

In the 1920s the colleges of the cubist painters encouraged such photographers as Laszlo *Moholy*-Nagy to experiment with photomontages. But by the 1930s the Frenchman Henri *Cartier-Bresson* had led fashion back to a more realistic technique: using the camera as an extension of the eye, he composed transient scenes in great detail. Picture magazines were also a feature of this period, one of the best-known being *Picture Post,* started in London in 1938 by a German refugee from Hitler, Stephen Lorant. Postwar photography has developed many of these themes, using the new precision cameras and techniques, and many galleries now have permanent exhibitions of photography.

Photography has also had a profound effect on painting. Some art historians believe that the 20th-centrury trend towards abstract painting has been primarily a reaction to the camera's ability to do superbly what representationalist painters have often only been able to do indifferently.

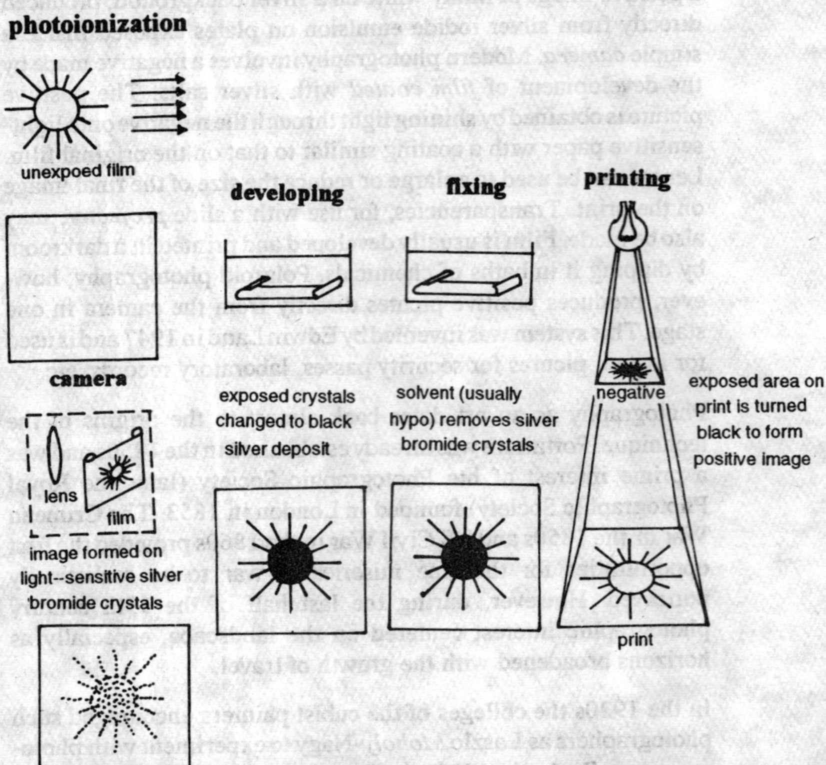

Photography. The stages in making a black and white photgraphic print.

Photometery. An optical instrument used for determining the illumination or luminous intensity of a source. The measurement, generally involves comparison of the given source with a standard source, both of which simultaneously illuminate a screen.

Photometry. (1) The branch of physics that deals with the measurement of the amount of light energy emitted or absorbed by a body. (2) The branch of physics concerned with measuring qauantities related to the intensity of light. These quantities are measured in two ways. If the intensity is measured in terms of the energy of the light, they are known as radiant quantities. They may also be measured in terms of their visual effect, since the sensitivity of the eye varies with the wavelength of the light. They are then known as luminous quantities.

Photomultiplier. An instrument that produces an amplified current when exposed to electromagnetic radiation. Photons striking a photocathode cause the emission of electons by the *photolectric* effect. These electons then strike another surface causing a greater number of elecrtrons to be emitted, the process being repeated several times to produce a greatly amplified current. Photomultipliers are used in certain *scintillation* counters.

Photon. A quanta of light. The energy of a photon of light radiation of frequency f is given by $hf = \frac{hc}{\lambda}$ where is λ its wave length.

Photovoltaic Effect. An *e.m.f.* develops between two layers of materials when an electromagnetic radiation is incident on them. This effect is known as photovoltaic effect.

Photosphere. The boundary between the atmosphere of a star and its much denser interior. The sun's photosphere is its visible surface, several hundred kilometres thick, lying between the opaque outer (convective) zone of the sun's interior and its transparent atmosphere. Almost all the energy emitted by the sun is radiated from the photosphere. Its temeprature falls from about 6000°C at the convective zone to about 4000°C where the photosphere merges with the chromosphere.

Physics. The study of the interreleationship between matter and energy, without reference to chemical change. Traditionally the subject was divided into the study of mechanics, electricity and magnetism, heat and thermodynamics, optics and acoustics. More modern aspects of the subject include quantum mechanics, relativity, nuclear physics, particle physics, solid-state physics and astrophysics.

Pi-mesons. Exchange particles which are responsible for strong nuclear force. Their charge can be positive, negative or zero. The mass of a π meson is approximatley 270 times the mass of an electron.

Pico. A prefix meaning 10^{-12}.

Pick-up. A device that converts information stored in a gramophone record into an electrical signal. In a record player, the pick-up cartridge, usually a removable assembly, consists of a stylus and a *transducer*. The stylus is forced to vibrate by the undulations in the record groove. the transducer responds electrically to these vibrations. The most common types of transducer use the *piezoelectric* effect or *electromagnetic* induction, but *capacitor*, variable-resistance, and variable-reluctance pick-ups are also used. A stereo pick-up has one stylus that picks up vibrations from both sides of the V-shaped groove. The components of the resulting two-dimensional vibration contributed by each side are at right angles and are separated by two suitably oriented transducers.

Piezo Electric Effect. The production of a small *e.m.f.* across certain substances on being subjected to external pressure. Such substances are known as piezo electric substances.

Pin Hole Camera. A closed box into which light can enter only through a very small opening. It forms an inverted image of an object placed in front of the opening on the wall opposite the opening. If a photographic plate is placed on the wall it forms a negative image of the object on it.

Pion. A group of three elementary particles classified as *mesons* (symbol:π) The charged pions π^+ and π^- have a mass of 139.6 MeV, the neutral pion (π^0) a mass of 136 MeV. The *strong* interaction can be represented by the exchange of virtual pions between particles.

Pitch. The characteristic of sound that depends on its frequency. A shrill sound is of a high pitch and corresponds to a larger frequency. Low pitch notes have low frequency.

Pitch of a Screw. The distance between successive crests of the thread in a screw, when measured parallel to the axis of the screw.

Pitot Tube. A type of *anemometer* for measuring the velocity of a fluid, invented by Henri Pitot (1695-1771). It consists of an L-shaped tube placed in the moving fluid; the vertical limb of the tube has an opening facing into the flow. The difference in pressure between the interior of the tube and the surroundings enables the velocity of the fluid to be calculated. The device is used to measure the velocity of liquids, aircraft airspeeds, etc.

Planck's Constant (h). A universal constant that connects the energy E and frequency f of a radiation E= *hf.* Its value is 6.625×10^{-34} joule second.

Planck's Formula for Black-Body Radiation. The energy radiated perunit area time at a given wave length λ, from a black body is given by the expression

$$E_v = \frac{2\pi hc^2}{\lambda^5} \frac{1}{\exp\left(\frac{hc}{KT}\right) - 1}$$

where c is the velocity of light, *h* is the Planck's constant, and T is the absolute temperature of the black body. *k* is the Boltzman's constant.

Planck Function. A thermodynamic function Y given by

$$Y = -(U/T) - (pV/T) + S$$

were U is internal energy and S entropy. It is equal to minus the quotient of the *Gibbs' Function* and the temperature T.

Planck's Law. Arising from *Planck's Quantum Theory,* Einstein postulated that the enegy content E of an energy packet or photon can be related to its frequency *v by*

$$E = hv.$$

This is often referred to as Planck's Law.

Planck's Quantum Theory. The Rayleigh-Jeans formula for the distribution of energy in the spectrum of black-body radiation is:

$$E = \frac{8\pi k_B T}{\lambda^4}$$

indicating a steady increase of energy density E with decreasing wavelenght λ as for the dotted curve in the figures. The experimental curve is shown by the full line and has a maximum for a given value of λ depending upon the temperature of the black body. This discrepancy was resolved by Planck, who showed that the error lay in applying the classical laws of mechanics to atomic particles.

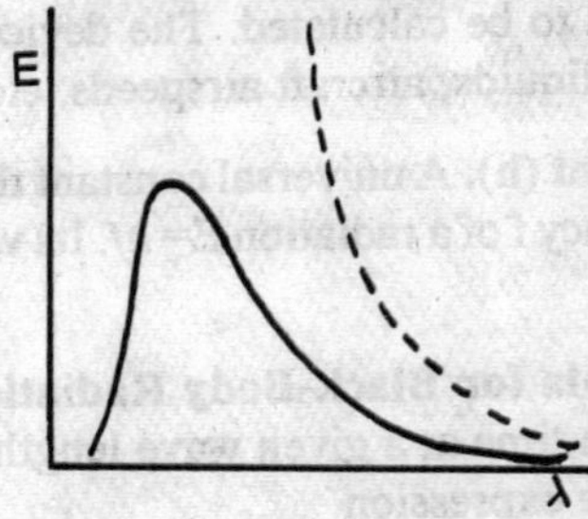

These observations were the basis of the modern quantum theory. In classical theory the amplitude of vibration of a linear oscillator may have any desired value within certain limits. Planck assumed, however, that the amplitude of vibration could only assume certain discrete values.

$$E = 0.\ h\nu,\ 2h\nu.....$$

Plane of Polarization. Refers to a plane perpendicular to the plane of vibration and containing the direction of propagation of light. It is also the plane containing the direction of propagation and the magnetic vector of the electromagnetic light wave.

Plane-Concave (Convex) Lens. A lens having one face plane and the other concave (convex).

Plante Cell. A primitive accumulator, consisting of rolled lead sheets dipping into dilute sulphuric acid.

Plasma. A gas the atoms of which have been completely ionized. Sometimes called the fourth state of matter, plasmas occur at enormously high temperatures, such as those in the sun. Plasmas are also created in *thermonuclear* reactors.

Plasticity. The permanent deformation of a substance after being sujected to a sufficiently large stress. Below a certain stress, called the elastic limit, most susbstances will recover their original shape when the stress is removed. Such substances are said to be elastic. However, single crystals of certain metals are plastic no matter how small the stress.

Pneumatics. The branch of physics that deal with the dynamic properties of gases.

Pockels Effect. If a light beam is passed through certain non-centrosymmetric crystals in the same direction as an applied electric field, the refractive indices of the crystals change with change of applied field, and the crystals become birefringent, the difference in velocity for the two rays being proportional to the electric field. If a plane polarized beam of light is used, the plane of polarization is rotated as the light passes through the crystal. The effect is analogous to the *Faraday Effect* in a magnetic field. The effect can be used in fibre-optical communications to modulate an incident laser beam.

Poise (P). A unit of dynamic viscosity. If the schearing stress per unit area τ in a liquid is expressed in dynes and the velocity gradient perpendicular to the direction of flow is du/dz, then the viscosity η defined by $\eta = \tau\,(du/dz)$ is in units of poise. 1 P = 0.1 N sm^{-2}.

Poise. A unit of viscosity in the *c.g.s.* system equal to the tangential force per unit area, in dynes per square centimetre, required to maintain unit difference in velocity, in centimetres per second, between two parallel planes in a fluid one centimetre apart. Named after J.L.M. *Poiseuille.*

Poiseuille's Formula. The volume of liquid V, flowing per second through a capillary tube of length l and radius r under a pressure p is given by

$$V = \frac{\pi p r^4}{8 l \eta}$$

where η is the viscosity of the liquid.

Poisson's Equation. It states that $\nabla . E = 4\pi\vec{\rho}$ where ρ is the volume density of charge and E is the electric field intensity.

Poission's Ratio. Ratio of fractional change in the cross sectional area. $\frac{\Delta A}{A}$, to the fractional change in the length $\frac{\Delta l}{l}$ of a material. It is a simple number.

Polarimeter. An optical instrument used to measure the angle of rotation of a plane-polarized light caused by an optically active susbstance. It consists of two nicol prisms—the first acts as a polarizer to produce a plane polarized beam from an unpolarised one; the plane polarized beam is passed through the optically active substance and the angle of rotation is measured with the help of the second nicol prism, known as the analyser.

Plarization (Polarized Light). Light is a transverse wave. When light travel's, the vibrations (of the electric field, say) take place in a plane perpendicular to the direction of propagation of the wave. In ordinary light, there are an infinite number of such directions in which these vibrations can take place. If however the vibrations are confined to a few directions only in this plane the light is a called partially-polarized light. If the vibrations are confined to one particular direction only, the resulting light is called plane-polarized.

Polarization, Angle of. The angle of reflection from a medium like glass, at which the reflected ray is completely polarized; the plane of vibration is at right angles to the plane of incidence.

Polarization Electric. When an insulator is placed in an electric field, the charges in its molecules separate in such a way that one face has a net Positive charge, and the other a net negative charge. The effect is known as electric polarization.

Polarizer. An arrangement or instrument that produces plane polarized light from unpolarized light.

Pole. Refers to the geometric centre of a spherical reflecting or refracting surface. All distances from such reflecting or refracting surfaces are measured from the pole.

Poles Magnetic. Points in a bar magnet where the magnetic forces are strongest. When a bar magnet is suspended so that it is free to oscillate, one of the poles points towards north and is known as North-seeking or North-or positive pole. The other that points towards South, is known as the South-or negative pole.

Pole Unit. A pole that when placed at a distance of one metre from a similar pole in vacuum, exerts a force of one newton on it.

Positive Column. A luminous region in an electric discharge in a gas near the positive electrode.

Positive Crystals. Doubly refracting crystals in which the ordinary ray travels faster as compared to the extra ordinary ray. e.g. Quartz.

Positive Lens. A lens having positive focal length and power, e.g. a convex lens.

Positive Rays. Refers to a stream of ions that are positively charged.

Positron. The antiparticle of the *electron*, having the same mass and *spin* as the electron but opposite electric charge. A positron and an electron annihilate each other on collision, producing two gamma-ray photons.

Positronium. Refers to an unstable assembly of a positron and an electron.It decays into photons.

Post Office Box. A device having resistances arranged in such a way that they can be used as a wheatstone bridge.It was used to find the resistances of long wires and thus to find any defects in the telephone or telegraph wires.

Potential, Electric. A measure of electrical work. The potential at a point in an *electric field* is one volt when one joule of energy is needed to bring a positive charge of one coulomb to that point from infinity. Usually the potential difference between two points, rather than the absolute potential, is used.

Potential Difference (V). Two points at different electric potentials are said to have a potential difference between them. When they are joined together with a wire, a current flows from the point at a higher potential to a point at a lower potential.

Potential Divider. Refers to an arrangement which is used for distributing the potential difference available from a source.

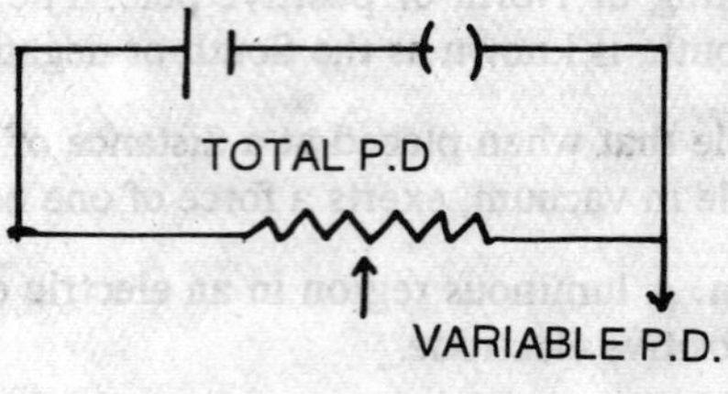

The entire potential difference may be applied across a rheostat, from which the required portion may be tapped making use of a variable (sliding) terminal.

Potential Energy. Energy possessed by a body by virtue of its position. The potential energy of a body of mass m raised through a height h is given by mgh.

Potential Energy of a Charged Conductor. E, is equal to $1/2\ QV = 1/2\ CV^2$ where Q is the charge on the conductor, V is the potential to which it is raised and C is the capacitance.

Potential Induced. The potential acquired by a body on account of the presence of another charged body in its neighbourhood. There is no transfer of charge to or from such a body.

Potentiometer. A variable electrical *resistance* in which a sliding contact can be adjusted to any position along its length so that a variable proportion of the resistance can be included in a circuit. Potentiometers are used for measuring unknown voltages and as a potential divider, producting a voltage output corresponding to calibrated positions of the sliding contact.

Pound. A unit of mass, the pound is defined as 0.453 592 37 kilogrammes.

Poundal. The *f.p.s.* unit of force. It is the force required to give a mass of one pound an acceleration of 1 foot per second 1 poundal = 0.138255 newton.

Power Photograph. A crystal diffraction method wherein the specimen is a randomly oriented crystalline powder.

Power. The rate at which a body or system does work. It is measured in watts or horsepower.

Power Factor. It is the ratio of the total power, in watts, dissipated in an electric circuit to the volt amperes applied to the circuit.

Power Station. An electricity generating plant that forms part of the *electricity* supply sustem. In *thermal power stations*, heat from the burning of fossil fuels (*oil, coal* and *gas*) or from thermal *nuclear* reactors is used to generate steam. The steam drives *turbines* connected to alternating-current generators (turboalternators), thus converting heat via mechanical energy into electricity. Gas turbines are also used, missing out the steam generation stage. They are easy to build and flexible to run, but have an efficiency of only about 25%. Sometimes the exhaust gas is used to help generate steam for a steam turbine, in order to increase the station's overall efficiency. Usually in a steam-turbine thermal station about 30 to 40% of the heat is converted to electricity, most of the rest being lost when steam is condensed to water before it is returned to the boilers. *Hydroelectric power stations* are more efficient (up to 90%) and provide a significant proportion of UK electricity supplies.

Power of Lens. Reciprocal of the focal length of a lens expressed in metres. Its unit is dioptre.

Power Reactor. A nuclear reactor designed to produce useful quantities of electric power.

Practical Units. Ampere, volt and ohm are known as the practical units of current, potential difference and resistance respectively.

Precession. When a body is rotating about an axis, it is said to precess if the axis of rotation moves around another axis at an angle to it.

Presbyopia. A defect of vision wherein the person is unable to see near objects, generally occurs in older people. It can be corrected by using convex lenses. *Also known* as long sightedness.

Pressure Force Per Unit Area. For a liquid, density d, the pressure at a depth h is hdg, where g is the *acceleration* of free fall. Pressure is usually measured in pascals, millimetres of mercury, or millibars.

Pressure Atmospheric (Normal). The pressure that will support a column of mercury 76 cm high at sea level at 0°C. It is equal to the pressure at a point at sea level on account of the weight of the air above it. Its value is nearly 10^5 N/m^2.

Pressure Gauge. An instrument for measuring the pressure of a fluid. Atmospheric pressure is measured with a mercury *barometer* or an aneroid barometer. Pressure above atmospheric pressure are usually measured with a Bourdon gauge, which consists of a flattened curved tube that tends to straighten under pressure. As it straightens it moves a pointer round a scale. Low pressures are measured with such vacuum gauges as the *McLeod gauge* or *Pirani gauge*.

Primary Cell. It converts chemical energy into electrical energy. The chemical reactions that take place in the cell are irreversible e.g. Daniel Cell.

Primary Colours. The minimum number of *colours* that, when mixed in the correct proportions, are capable of giving all the other colours in the visible spectrum. When light of three *primary additive colours* (usually red, green, and blue) is mixed in equal intensities, white light results. This principle is used in *colour television* and *colour photography*. Any particular colour can also be obtained by subtracting from white light a mixture of three *primary subtractive colours*, usually cyan (blue-green), magenta (purplish red), and yellow, which are complementary to red, green and blue. Adding pigments of these in equal proportions gives black pigment.

Principal Axis. Refers to the line joining the centres of curvatures of the surfaces of a lens.

Principal Planes. A pair of conjugate planes perpendicular to the principal axis, that correspond to unit positive transverse magnification. If an incident ray cuts one of the principal planes at a height *h* have above the axis, the emergent ray cuts the other principal plane at the same height (h).

Principle of Moments. For any system in equilibrium, the sum of the clockwise moments is equal to the sum of the anti-clockwise moments.

Principle of Superposition. When two identical waves pass through a given region, the amplitude at any point in the region is equal to the algebraic sum of the amplitudes due to each wave.

Printed Circuit. An electronic circuit in which the connections between components are formed by a pattern of conducting film on a board, instead of by wires. The method greatly facilitates mass-production. An insulating board is coated with a conducting material, such as copper, and a protective pattern is deposited on it using photographic techniques. The unprotected metal is then etched away and components are soldered in place. In television receivers, the replacement of thermionic valves by plug-in printed- circuit boards has made maintenance easier. A faulty circuit board can be replaced without tracing the failed component.

Prism. A piece of glass or other transparent material having parallel polygonal ends, with a number of rectangular surfaces meeting them at right angles. They are used in optical instruments, such as cameras and binoculars, for changing the direction of light, either by refraction or by reflection from its walls. They are also used for splitting light into its component colours by double refraction, once on entering the prism and again on leaving it.

Prism Table. A circular base having three levelling screws. It can be brought to a range of desired heights and clamped at any height in this range. It can rotate about an axis passing through its centre. It carries a circular scale that can be used to read the position of the prism or grating placed on it.

Progressive Wave (also called Travelling Wave). A wave motion wherein the whole periodic displacement advances through the medium. At any instant, the disturbance changes with time and is represented by the equation

$$y = a \sin 2\pi \left(\Omega - \frac{x}{\lambda} \right)$$

where a is the amplitude, Ω is the frequency and λ the wave length of the wave.

Prompt Neutron. A neutron produced in a reactor by primary fission and not by decay of fission product.

Projectile. A body thrown into space. If a projectile is thrown with a velocity V at an angle θ to the horizontal, the maximum horizontal distance covered by the body $= \frac{V^2 \sin 2\theta}{g}$. Its total time flight $= \frac{2V \sin 2\theta}{g}$, and the maximum height upto which it rises $= \frac{V^2 \sin^2 w}{g}$.

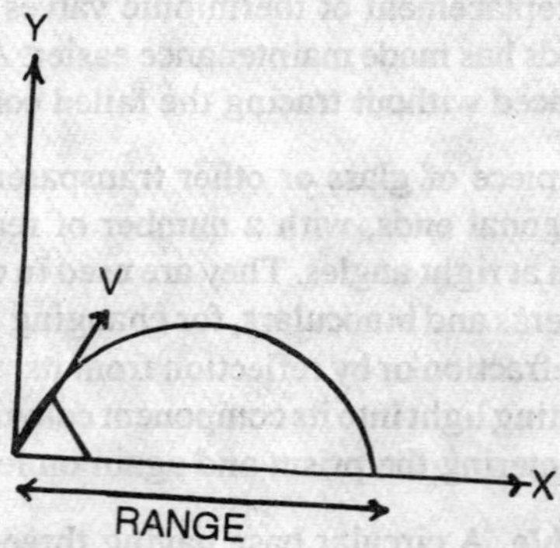

Projection, Angle of. The angle between the velocity of projection of a body and the horizontal.

Propeller. A device for converting the rotation of a shaft into thrust in the direction of its axis. A *marine propeller* is attached to a shaft connected to the ship's engine and has between two and six blades, shaped as part of a helical surface. It acts as a screw, accelerating a column of water rearwards. An *air popeller* (or airscrew), moving in a thinner medium, has longer thinner blades (usually two or four) and a higher rotational speed. It, too, accelerates a mass of air rearwards thrusting the aircraft forwards by reaction.

Proper Motion. The rate at which a star is observed to move on the *celestial* sphere, i.e. perpendicular to the line of sight. Values are very small, often negligible.

Proportional Counter. A device in which the number of ions produced by an ionzing radiation is counted by measuring the height of the electric pulse produced by it.

Proton. An elementary particle which is present on nuclei of all atoms. Its mass is 1.67×10^{-27} kg and its charge 1.6×10^{-19} coulomb.

Proton Number (also called Atomic Number) (Z). The number of protons present in the nucelus of an atom. It is equal to the number of electrons in an atom.

Proton-Proton Cycle. A chain of nuclear fusion reactions supposed to be responsible for production of energy in the Sun. The reactions lead to the conversion of hydrogen into helium and are given by

$$_1H^1 + {_1H^1} \rightarrow {_1H^2} + \nu + e^+$$

$$_1H^2 + {_1H^1} \rightarrow {_2He^3} + r$$

$$_2He^3 + {_2He^3} \rightarrow {_2He^4} + 2\,{_1H^1}$$

Psi Particle. An elementary particle discovered in 1974 and classified as a meson. Its unusually long lifetime led to the extersion of the quark model (*see* particle physics) to include a quark (and its antiquark) having a new property called *charm*. The psi particle itself is not charmed as it consists of a charmed quark and its antiquark.

Psychrometer. A hygrometer wherein strong draught is obtained past the wet and dry bulbs either by whirling on a sling or by a fan.

P-type Conductivity. Conductivity in a semiconductor due to flow of holes.

P-type Semiconductor. An extrinsic semiconductor wherein the density of mobile holes is more than that of conduction electrons.

Pulley. A simple machine consisting of a wheel with a grooved rim in which a thread or rope can run and a frame in which the wheel turns freel on an axle. In the fixed pulley the frame is attached to a rigid support where as in the movable pulley the frame is free to move. By combining fixed and movable pulleys, a small effort can be used to overcome a large resistance.

Pumps. Machines designed to transfer mechancial energy to a fluid. This energy may be required to move the fluid from one place to another, raise its level, increase its pressure, or provide circulation. Pumps are of three kinds: reciprocating, centrifugal, and rotary displacement. The reciprocating pump consists essentially of a piston moving to and fro; liquid is drawn in through one valve on the down stroke, and expelled through an exhaust valve on the upstroke. The centrifugal pump draws liquid into the centre of a rotating impeller, which thrusts it out of the exit. Rotating pumps consist of shaped gears or rotors, which move the liquid around a close-fitting chamber.

Pupil. Refers to aperture of the eye. It is adjustable in size and generally appears black.

Pycnometer. A type relative density (specific gravity) bottle consisting of a bulb between two capillary tubes, on one of which is reference mark. The density of a liquid is measured by finding the weight of the known volume filling the vessel up to the mark.

Pyrometer. An instrument for measuring very high temperatures, by measuring the radiation emitted by the body.

Pyrometry. Measurement of high temperatures making use of pyrometers.

Q

Quadripole. An elecrtrical network which has four terminals only i.e., a pair of input terminals and a pair of output terminals. If its electrical properties remain unaffected when the input and output terminals are interchanged, it is called as a *symmetrical quadripole.*

Quality of Sound. Most musical notes to contain more than one frequency. Quality of sound is the characteristic of a musical notes which depends on the frequencies present in the note. Each note generally consists of a fundamental frequency, having highest intensity, and several overtones, whose frequencies are integral multiplies of the fundamental and whose intensity is much lower. The quality changes with the number of overtones present and their intensity.

Quantum. A packet of energy : according to quantum theory, the amount of energy released or absorbed in a physical process is always an integral multiple of discrete unite of energy, known as quantum (of energy). The quantum of electromagnetic radiation is known as a *photon.*

Quantum Electronics. Deals with application of the principles of quantum mechanics to study of the production and amplification of power at frequencies from 10^9 to 10^{11} hertz in solid crystals.

Quantum Mechanics (Quantum Theory). A theory based on the idea that most physical quantities, such as energy, angular momentum, etc., can take up only certain discrete values and cannot change continuously. The theory was first used successfully by Max Planck (1900) to explain the black body radiation from hot bodies. The theory was later used to explain the photoelectric effect, compton effect and developed by Niels Bohr to account for the hydrogen spectrum successfully. Quantum mechanics, developed from quantum theory, is a system of mechanics which explains very successfully the physical phenomena that occur on a very small scale, such as the motion of nuclear particles and electrons in the atom. These phenomenon cannot be explained on the basis of the Newtonian Mechanics.

Quantum Numbers. Integral or half-integral numbers used to specify the state of system in quantum mechanics. e.g., the electrons moving around the nucleus of an atoms is completely specified by the four quantum numbers, *n* (principal quantum number), *l* (azimuthal quantum number). *m* (magnetic quantum number) and *s* (spin quantum number). They can respectively take the values (1,2,3,...), (0,1,2,......, *n*-1), —(—l,—l+1,0 1,2,...l) and ± 1/2.

Quark. Hypothetical fundamental particle (s) that are postulated to be building blocks of elementary particles.

Quartz. It is a double refracting crystal, optical uniaxial and positive and rotating the plane of polarization to the left or right according to the variety, and to different extents for different colours. It transmits wavelengths between 180-4000 nm.

Quasar (Quasi-Stellar Object; or QSO). A class of celestial objects disco- vered in 1964-65, lying beyond our Galaxy. They appear as starlike points of light, each emitting more energy than several hundred giant galaxies. Quasar *redshifts* are extremely large indicating that they are the most distant and the youngest extragalactic objects known. They must be highly luminous to be visible at such distances. About 3500 quasars are known. The energy-producing region is extremely compact; the prodigious energy could result from matter spiralling into a supermassive *black hole* (may be 10^9 solar masses) at the centre of a galaxy.

Quiescent Current. The current flowing in the circuit under conditions of zero applied signal

Q. Value. Amount of the energy given out in a nuclear reactor. It is generally expressed as Me V/reaction.

R

Rad. The unit of absorbed dose of ionizing radiation in the *c.g.s.* system equal to an energy absorption of 100 ergs per gram of irradiated material.

Radar (Radio Detection and Ranging). A method of detecting distant objects based on the reflection of microwaves (10-1000 mm wavelength). It was develeoped during the years preceding World War II by a British team led by Sir Robert Watson-Watt as a means of detecting enemy aircraft. In primary radar, a high-powered beam of pulses is transmitted from a rotating aerial and reflected

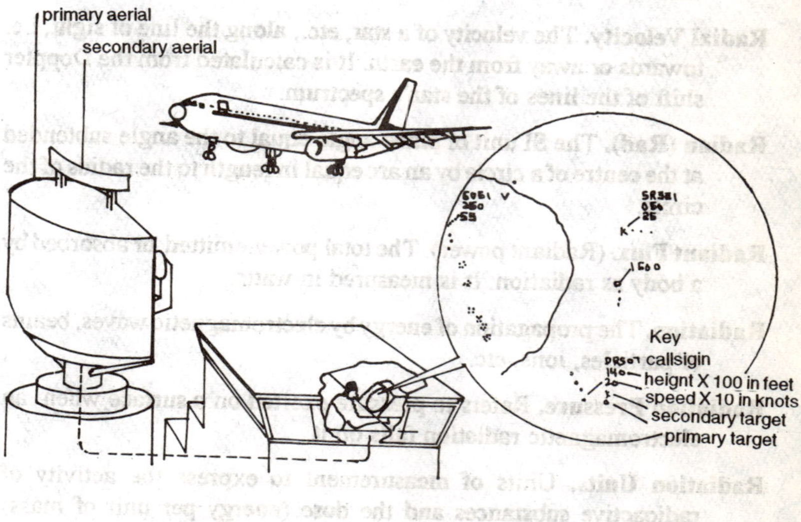

Radar. The high-frequency pulses sent out by the transmitting aerial are reflected back by the target and diplayed on a screen. The typical air-traffic control picture seen here shows East Anglia (only a few of the targets are labelled, for clarity).

back to the same aerial by any object it encounters. Various computerized techniques are used to filter out unwanted reflections, an image of the target (aircraft, missile, etc.) being displayed on the screen of a *cathode-ray tube* (CRT). The beam of the CRT scans the screen of the tube presenting a plot of the target area and any targets within it. Radar is the basis of air-traffic control and can detect targets at 450 km to an accuracy of about 100 metres. Primary radar is also carried on aircraft themselves for mapping and weather detection. It is used in sea navigation and by the police to detect speeding motorists (using the *Doppler effect*).

In secondary radar, coded pulses from the ground are sued to obtain a reply from a *transponder* carried by aircraft, which signals back various information, including the aircraft's call sign and altitude. This information is displayed on the radar screen, together with an image of the target, using letters and numbers. Computer processing enables the diplay to include targets from several different radar transmitters.

Radial Velocity. The velocity of a star, etc., along the line of sight, i.e. towards or away from the earth. It is calculated from the Doppler shift of the lines of the star's spectrum.

Radian (Rad). The SI unit of plane angle equal to the angle subtended at the centre of a circle by an arc equal in length to the radius of the circle.

Radiant Flux. (Radiant power). The total power emitted or absorbed by a body as radiation. It is measured in watts.

Radiation. The propagation of energy by electromagnetic waves, beams of particles, ions, etc.

Radiation Pressure. Refers to pressure exerted on a surface when an electromagnetic radiation falls on it.

Radiation Units. Units of measurement to express the activity of radioactive substances and the dose (energy per unit of mass) absorbed by matter exposed to *ionizing* radiation The *curie, roentgen,* and *rad* have been replaced by the SI units *becquerel, gray* and *sievert.*

Radiative Collision. A collision which takes place between two charged particels and from which electromagnetic radiation is emitted.

Radio. The transmission of information by radio-frequency (3 kilohertz to 300 gigahertz) electromagnetic waves. The use of radio waves for communication was pioneered by *Marconi* in 1859, although their existence was postulated by *Maxwell* in 1873 and demonstrated by *Hertz* in 1888.

A *radio transmitter* generates a radio frequency electrical signal (carrier wave) in an oscillator and superimposes the sound signal on it by modualtion. The composite signal is then fed to an aerial, which transmits electromagnetic waves. Although the waves travel in straight lines, transmission beyond the horizon is made possible by the ionosphere, a layer of the upper atmosphere that reflects long and medium waves. Long-wave (30 to 300 kilohertz) and medium-wave (300 to 3000 kilohertz) radio is used for direct (ground-wave) transmission in amplitude-modulated (AM) sound broadcasting. Short-wave radio is used for AM broadcasting over longer distances using the ionosphere (sky waves) and for communication to ships and aircraft. VHF (very high frequency; 30 to 300 megahertz) is used for *stereophonic* sound broadcasting in frequency modulated (FM) transmissions. Television uses mostly UHF (ultrahigh frequency). Neither VHF nor UHF are reflected by the ionosphere, but can be relayed by *communications* satellites. All *radio receivers* consist of an *aerial,* which detects the signal and converts it to an electrical current; a resonant circuit, which selects the required frequency band; a demodulator to retrieve the audio-frequency signal from the modulated carrier wave; an amplifier; and *loudspeakers* to reproduce the sound.

Radioactive Age. The age of a geological or an archeological specimen as calculated by radioactive dating.

Radioactive Collision. A particular type of nuclear reaction wherein a neutron enters a nucleus and a γ–photon is emitted. The new nucleus formed has the same atomic number but its mass number is increased by unity. This types of reaction is used for production of radioisotopes.

Radioactive Dating. A method of measuring the age of a specimen making use of the principles of radioactivity. *See* carbon dating.

Radioactive Series. A series of isotopes each of which after the first is the dacay product of the preceding one. The last member of the series, is stable.

Radioactive Tracer. Refers to definite quantity of radioisotope introduced into a biological or mechanical system so that its path through the system and its concentration in particular areas can be determined by the meauring the radioactivity.

Radioactive Waste. Solid liquid, and gaseous waste products from nuclear reactors, uranium processing plants, hospitals, etc., that are radioactive. Because the radioactivity of some materials will remain for thousands of years, their disposal must be undertaken with great care. High-level waste (spent nuclear fuel, etc.) needs artificial cooling and is therefore stored by its producers for several decades before disposal. Intermediate-level waste (reactor components, filters, sludges, etc., from processing plants) is solidified and stored mixed with concrete in steel drums in power stations prior to burial in deep mines or beneath the seabed in concrete chambers, where it cannot contaminate ground water. Low-level waste (solids and liquids contaminated by traces of radioactivity) presents fewer problems; in the UK, since 1988 it has been disposed of by a special company (Nirex Ltd), set up by the government and the nuclear industry, in steel drums in concrete-lined trenches at Driggs, Cumbria. Other countries make similar arrangements. Until 1983, when it was suspended by international agreement, low and intermediate-level wastes were disposed of in the deep Atlantic in steel drums cast in concrete. In addition some very dilute low-level gaseous and liquid wastes have been discharged into the air and the sea.

Radioactivity. The decay of certain naturally occuring nuclides with the emission α-, β- and γ- rays. Radioactivity when induced in certain stable nuclei by bomarding them with high energy particles is known as *artificial radioactivity*.

Radio Astronomy. The study of celestial objects by means of the radio waves they emit. This radio emission was first noticed by K. G. *Jansky* in the USA in 1932 and developed in the UK during and after World War II, especially at *Jodrell Bank* and *Cambridge University*. It now forms an important branch of *astronomy*.

Radiobalance. An instrument wherein the heating due to the absorption of radiation is neutralized by the cooling due to the Peltier effect at one in junction of a thermocouple.

Radio Frequency. Refers to the frequency of electromagnetic radiation in the range 10 kilohertz to 1,00,000 megahertz.

Radiology. Refers to the study and application of X-rays, gamma rays, and other penetrating ionizing radiation.

Radioluminescence. The process of the emission of visible electromegnetic radiation from a radioactive substance.

Radiolysis. The process of the chemical decomposition of material into ions, excited atoms and molecules, etc., by ionizing radiation.

Radiometer. An instrument for measuring radiation in energy or power units. The term is generally used for instrument that detect and measure infrared radiation.

Radiomicrometer. A very sensitive detector of radiation. A thermocouple of antimony and bismuth is joined in series with a single copper loop suspended by a quartz fibre between the two poles of a magnet. Radiation falling on one Sb-Bi junction causes a current to flow in the copper loop which is thus deflected, the deflection being measured from the movement of a spot of light reflected from a mirror carried by the loop.

Radio Receiver, One Valve. Simplest type of radio-receiver using one triode valve, that acts as a detector as well as an amplifier. The circuit diagram of such a receiver is given in the figure. The tuning is carried out with the help of the variable capacitor C. The output of such a receiver is not very high.

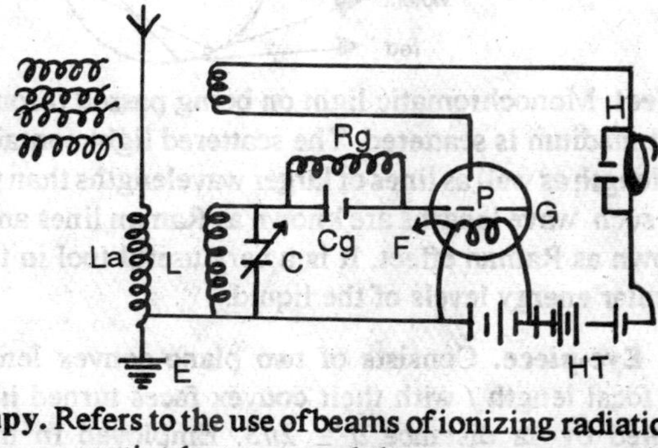

Radiotherapy. Refers to the use of beams of ionizing radiations, such as X-rays, energetic electrons, and the stream of gamma rays, from the radioisotope cabalt-60 in the treatment of cancer. A sufficient amount of radiation must be given to kill the cancer cells without harming intervening tissue. This is achieved by irradiating from several different directions with a narrow beam so that the tumour receives the maximum dose.

Radio Waves. Electromagnetic waves having wavelengths greater than a few millimetres. They are used for carrying information and for broadcasting educational and entertainment programmes from a broadcast centre also used for television transmission.

Radius of Curvature. (R) The radius of the sphere of which a mirror or lens surface is a part.

Radius of Gyration (k). Refers to the distance from the axis of rotation of a body to a point where the whole mass may be supposed to be concentrated. It is given by

$$k = \frac{I}{m}$$

where I is the moment of inertia of the body about the axis of rotation.

Rainbow. An optical phenomenon consisting of an arc of light across the sky composed of the colours of the spectrum. It is caused by the refraction of sunlight through falling water drops; the larger the drops the stronger the colours.

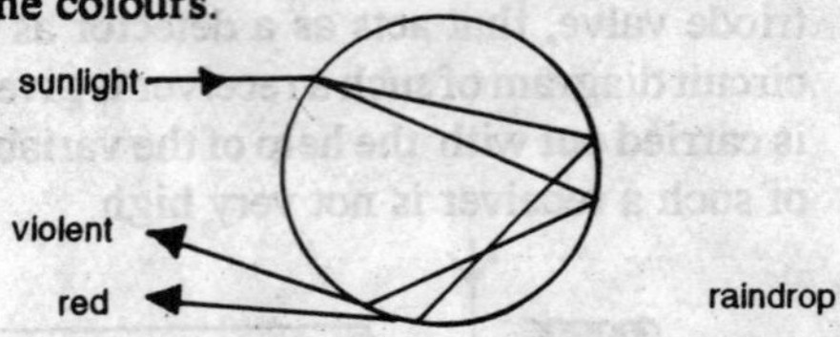

Raman Effect. Monochromatic light on being passed through a transparent medium is scattered. The scattered light contains original wave length as well as lines of larger wavelengths than the original lines, such wave lengths are known as Raman lines and the effect is known as Raman effect. It is a very useful tool in the study of molecular energy levels of the liquid.

Ramsden's Eye-piece. Consists of two plano-convex lenses having equal focal length f with their convex faces turned inwards and separated by a distance $d = 2f/3$. Employed in most of the measuring optical instruments.

Rankine Temperature. Symbol; °R. An obsolete thermodynamic temperature scale linked to the Fahrenheit degree. Absolute zero on this scale is -459.67°F. °R = °F + 459.67. The ice point is 491.7°R, generally taken as 492°R. 1 k =1.8 R.

Rarefaction. Low pressure and low density regions in a medium obtained during the propagation of longitudinal waves the distance between two successive rerefactions gives the wavelength of the wave.

Ray. The ractilinear path adopted by a narrow beam of light on any other type of radiation.

Rayleigh's Law. In magnetic materials subjected to magnetic forces low in comparison with the maximum coercive force, the hysterisis loss in a cycle changes directly as the cube of the induction.

Reactance (X). It is measure of the resistance offered by an inductance or capacitance to an alternating current of freaquency *w*. The reactance of an inductor of inductance L is given by $X_L = L_w$ while for a capacitance, it is given by

$$X_e = \frac{1}{cw}$$

Reactive Load. The load wherein the current and the voltage at the terminals are out of phase with each other.

Reactor. An electrical apparatus possessing reactance used because of that property.

Rectifier A device that allows electric current to flow in one direction only; it is generally used to convert alternating current to direct current. *Semiconductor* diodes have now replaced thermionic valves (diodes) for lower voltage applications. For power supplies of several megawatts, mercury—arc rectifiers are used.

Red Gaint. A type of cool gaint star that emits light in the red region of the spectrum. A star expands to a red gaint when it exhausts its nuclear fuel.

Redshift. An overall displacement towards large wavelengths of the spectral lines of a celestial object. Its astronomical significance was suggested by Edwin *Hubble* in 1929, when it was used as the basis of the theory that the universe is expanding. A redshift usually arises from the *Doppler* effect, that is from recession of a celestial object. It increases as the object's radial velocity increases and for an extragalactic body can be used as a measure of distance. A *gravitational redshift* occurs whenever radiation is emitted by a body: it is generally negligible except when the gravitational field is very strong.

Reflectance. It is the ratio of the reflected light to the incident light on a surface.

Reflecting Power. The ratio of the amount of energy reflected to the amount of energy falling on a body per unit time is known as its reflecting power.

Reflecting Telescope. A telescope which makes use of a concave mirror as an object glass. Reflecting telescopes are free from various defects such as chromatic aberration etc., of refracting telescopes.Most of the large astronomical telescopes belong this category.

Reflecting Angle of. Angle of a reflection

Reflecting, Laws of. (1) The angle of incidence and the angle of reflection are equal. (2) The incident ray, the reflected ray and the normal all lie in the same plane.

Reflection. The rebounding of a wave of light or other radiation when it strikes a surface. Reflected light obeys two laws: first, the normal (an imaginary line vertical to the surface at the point of impact), the incident ray, and the reflected rays all lie in the same plane; second, the angle of incidence (i.e., the angle between the incident ray and the normal) is equal to the angle of reflection.

Reflection of Wave. When waves are reflected at a denser medium they undergo a phase change of π. In phase the reflection takes place at a rare medium no such change of phase of the wave occurs.

Reflectivity. Symbol ; α *p.* The reflectance of a layer of material sufficiently thick that no change of reflectance would take place with increase in thickness.

Reflector. Refer to a layer of material surrounding the core of a nuclear reactor, its purpose is to reflect some of the escaping neutrons back into the core.

Reflection Total Internal. Total internal reflection.

Refraction. The bending of a beam of radiation as it passes from one medium onto another. For a light ray the amount by which it is bent depends on the angle of the incident ray and on the refractive indices of the two media, the exact dependence being given by *Snell's law.* Refraction is caused by the difference in the velocity of the radiation in the two media. The ratio of the velocity of light in the two media is known as the *refractive index.* If the first medium is a vacuum, the ratio is known as the absolute refractive index of the second medium. The refractive index of glass lies between 1.5 and 1.7, for diamond it is 2.1, and for water 1.33 (at 25°C).

Refraction, Laws of. (1) The incident ray, the refracted ray and the normal all lie in the same plane, (2) The ratio $\frac{\sin i}{\sin r}$ = Constant, where *i* and *r* are the angles of incidence and refraction respectively. The second law is also known as Snell's law.

Refractive Index. When a ray goes from one medium to another medium, the ratio $\frac{\sin i}{\sin r} = \mu$, is a constant and is known as the refractive index of the second medium with respect to the first medium.

Refractometer. A device used for measuring the refrative index of a medium.

Regnault's Apparatus. An apparatus for measuring the absolute coefficient of thermal expansion of a liquid.

Regulation. The changes that take place in the available voltage due to internal resistance or to internal impedance when the load is varied under specified conditions.

Rejector. The impedance of a circuit made up of inductance and capacitance in parallel has a maximum value at some particular frequency. Such a circuit is a rejecter for that frequency and its maximum impedance is known as the *rejector impedance* or *dynamic impedance*.

Relative Density (Specific Gravity). The density of a substance as compared to the density of water at 4°C.

Relative Humidity. It is the ratio of the amount of water vapour present in the air to the amount of maximum amount of water vapour the air could hold at that temperature.

Relative Permeability. (εr): The ratio $\frac{C_m}{C_0}$ where C_m is the capacitance of a parallel plate capacitance placed in the medium and C_O the capacitance in air or vacuum.

Relative Velocity. The velocity of an object A with respect to another object B. If V_A, V_B denote the velocities of A and B respectively, the relative velocity of B with respect to A would be $V_B - V_A$ if both are moving in the same direction.

Relativistic Mass. The mass of a body that is moving at a velocity comparable to the velocity of light. According to the theory of relativity, if the velocity of the body is v then its mass is $m_0 (1 - v^2/c^2)^{-1/2}$ where c is the velocity of light and m_0 the rest mass (the mass when stationary). The relativistic mass of an electron travelling at 99% of the velocity of light is seven times its rest mass.

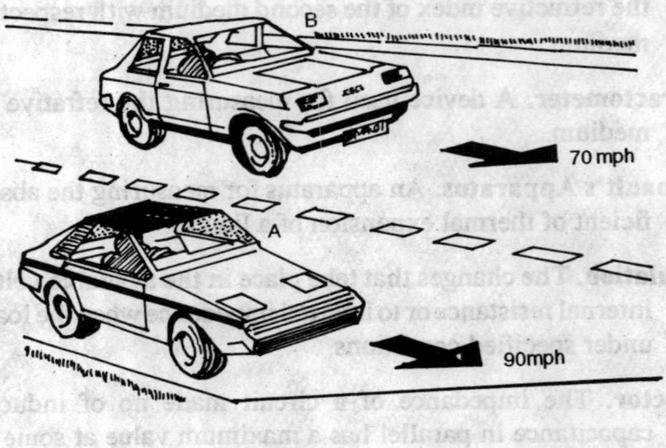

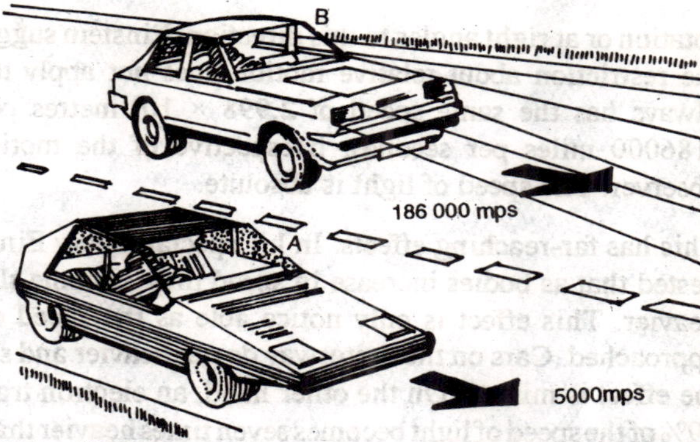

Relativity (1) Two cars, A and B, travelling along a motorway. The speed of car A relative to the earth is 90 mph. The speed of car B relative to the earth is 70 mph. The speed of car A relative to car B is 90-70=20 mph. 2. Car B is now stationary; it headlight beams are turned on as supercar A passes, travelling at 5000 miles per second. The speed of car B's headlight beam relative to car A is 186000 miles per second (not 186 000-5000 miles per second). The velocity of light is absolute.

Relativistic Particle. A particle moving with a velocity close to the velocity of light, say greater than $0.1c$, c being the velocity of light.

Relativity. An important theory proposed by Albert *Eingstein*. The first part of the theory, published in 1905 and known as the special theory, applies only to motion in which there is no acceleration. The problem that Einstein set out to solve was concerned with the speed at which light travels relative to an observer.

Up to this time it was thought that light travelled through a stationary medium, called the ether, at a constant speed and that its speed relative to an observer could be calculated in the same way as the relative speed of any two moving objects. For example, if one car (A) travelling at 90 mph on a motorway overtakes another (B) travelling at 70 mph, the speed of the two cars relative to each other is 20 mph. In talking about relative speeds it is necessary to be precise about what a particular speed is relative to. Car A's speed relative to car B is 20 mph, but relative to the earth it is 90 mph. Relative to the sun it is about 64 000 mph. Because two American scientists, Michelson and Morley had shown that light travelled at the same speed whether measured in the direction of the earth's rotation or at right angles to this direction, Einstein suggested that the restriction about relative motion does not apply to light. It always has the same speed of 2.998×10^8 metres per second (186000 miles per second), irrespective of the motion of the observer. The speed of light is absolute.

This has far-reaching effects. In his special theory Einstein suggested that as bodies increase in speed they become shorter and heavier. This effect is only notice able as the speed of light is approached. Cars on the motorway do get heavier and shorter but the effect is minute. On the other hand, an electron travelling at 99% of the speed of light becomes seven times heavier than its mass at rest. Einstein aslo showed that no body could actually travel at the speed of light itself. If it did it would be infinitely heavey and have zero length.

The increase in mass and decrease in lenght that a body undergoes when moving at high speeds led Einstein to the conclusion that mass (m) and energy (E) are two different aspects of the same thing; they are related by the simple equation $E = mc^2$, where c is the velocity of light. The atom bomb and *nuclear* energy both depnd on this equation.

In the general theory, of 1916, Einstein considered accelerated relative motion, especially as it is concerned with gravitation. In this theory, the gravitational force experienced by a body is treated as a property, of space and time, which Einstein suggested was "curved" by the presence of the mass. The motion of the stars and planets is controlled by this curvature of space in the vicinity of matter. Light, too, is bent by the gravitational field of a massive body. The observed bending of light rays as they pass close to the sun and the shift of certain lines in the solar *spectrum* have provided experimental verification of the theory.

Relay, Electric. An electrical switching device in which one circuit is controlled by a separate circuit, often to avoid the danger of direct contact with high-voltage supplies or to reduce the length of high-current cable needed. Relays may involve electromagnetic connection between the two circuits or be controlled by *solid-state* components.

Reluctance (or Magnetic Resistance). The ratio of the *magnetomotive* force in a material to the total *magnetic flux* induced by it. It is analogous to electrical *resistance*.

Resistance Box. Set of combination of resistance coils arranged in such a manner that any desired value of resistance can be obtained by using one or a combination of these. In one of the most widely used forms, each resistance coil possesses its ends soldered to stout brass pieces fixed to an ebonite plate that form the top of the reistance box. The brass pieces are separated from each other by small spaces filled with brass plugs. If the space is closed with the plug, the current flows through the brass pieces and effective resistance is zero. When the plug in open, the current starts flowing through the coil and the resistance of the coil comes into the circuit.

Resistance, Electrical. Refers to the ratio of the potential difference across the ends of a conductor to the current flowing through it. For a conductor, it has been found to be directly proportional to the length and inversely to the area of the conductor.

Resistance Thermometer. A thermometer based on the principle of change of resistance of a conductor with temperature. The variation of resistance with temperature is given by the relation $R_t = R_o(1 + \alpha t)$ when R_0, R is the constant the resistance at 0 degree and t degree and α is a constant.

Resistance in Parallel. A number of resistance are said to be connected in parallel when one end of all is joined to one point and other end is joined to another common point. If resistance R_1,...R_2,...R_n are joined in parallel, their equilvalent resistance across AB, is given by

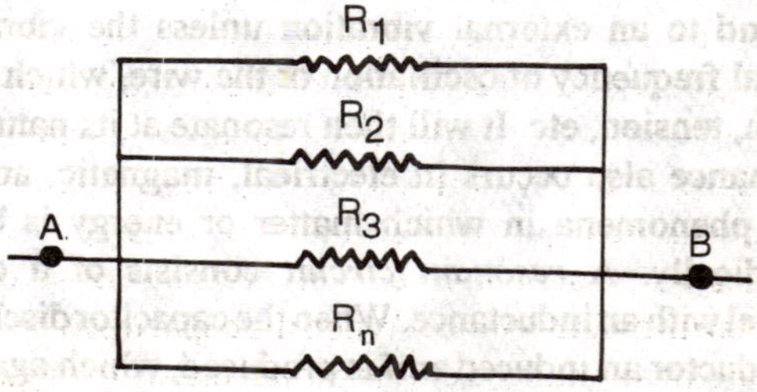

$$\frac{1}{R} = \frac{1}{R_1} + \frac{1}{R_2} + \frac{1}{R_3} + \ldots\ldots \frac{1}{R_n}$$

Resistance in Series. A number of resistance are said to be connected in series when one end of one is connected to one end of second,

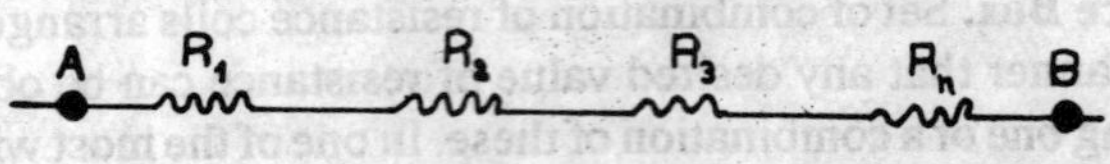

second end of second is connected to one end of the third,..... and so as shown in the figure.

If resistance R_1, R_2.... are connected in series the total resistance R is given by

$$R = R_1 + R_2 + \ldots\ldots + R_n.$$

Resistivity (e) (Specific Resistance). The resistance of a material of unit length and of unit area of cross section ρ is called the resistivity of material. It is measured in ohm-metre ($\Omega\, m$).

Resolution of Vectors. The division of vector into components, generally at right angles to each other.

Resolving Power (R.P.). The measure of the ability of an optical instrument to form separate and distinguishable image of two objects very close to each other. For telescope, the resolving power is given by $1.22 \dfrac{\lambda}{a}$ where λ is the wave length of light used and a is the aperture. For a grating the R.P. = n N where N is the number of lines in the grating and n is the order of the spectrum.

Resonance. The sympathetic oscillation of a system in response to an external excitation. A wire under tension, for example, will not respond to an external vibration unless the vibration is at the natural frequency of oscillation of the wire, which depends on its length, tension, etc. It will then resonate at its natural frequency. Resonance also occurs in electrical, magnetic, and electromagnetic phenomena in which matter or energy is being changed periodically. A *resonant circuit* consists of a capacitance in parallel with an inductance. When the capacitor discharges through the inductor an induced emf is produced, which again charges the capacitor (in the opposite sense). The circuit will then continue to oscillate, provided that energy is supplied from an outside source, at a frequency determined by the values of the capacitor (which is usually a variable device) and the inductor. Such circuits, feeding an aerial, are used in *radio transmitters* to generate radio-frequency oscillations, and in receivers to detect incoming signals.

Resonance Circuit. An electric circuit having an inductance and capacitance, when supplied current from an alternating current source of frequency

$$f = \frac{1}{2\pi\sqrt{LC}}$$

where L and C are the inductance and the capacitance.

Restitution Coefficient of. The ratio of the relative velocity of two colliding bodies immediatley after collision to their relative velocity before collision.

Rest Mass (m_0). The mass of an object as measured by an object at rest with respect to the observer.

Rest Energy. The energy equivalent of the rest mass of a body generally expressed in electron volts.

Resultant. A vector that has the same effect as a number of vectors, is known as the resultant of all these vectors.

Retardation (Decleration). Negative acceleration.

Retina. The light-sensitive layer that forms the inner surface of the eye. Light is focused by the lens onto the retina, which contains numerous interconnecting light-sensitive cells that send signals to the visual centre of the brain via the optic nerve. These cells are of two kinds: the rods, which are sensitive to light and dark and are responsible for vision in dim light; and the cones, which are sensitive to colour and visual detail. The greatest concentration of cones occurs at a small area at the back of the eye (the fovea), which is therefore the area of greatest visual acuity.

Reverberation. Refers to persistance of sound even after the source is not emitting any sound.

Reversible Changes (Reversible Process). An ideal process wherein the system undergoing the change remains at equilibrium throughout. Such a process can be reversed such that all the changes constituting the process can occur in the reverse direction. For a process to be reversible, it would have to be carried out infinitely slowely without involving any loss of energy due to friction etc. No real process taking place in nature is, reversible.

Rheology. Refers to the study of the deformation and flow of matter.

Rheometer. An instrument used to measure the flow properties of solid materials by investigating the relationship between stress, strain and time.

Rheostat. A variable resistor generally a coil would over a cylindrical block. The resistance is changed by sliding a contact on the coil.

Right Hand Rule. *See* Fleming's right-hand rule.

Rigid Body. A body in which the distance between every pair of particles remains constant under the action of force.

Rigidity Modulus. *See* elasticity modulus.

Rocket. Any of several annual or biennial herbaceous plants of the genus *Sisymbrium,* having yellow flowers. London rocket (*S. irio*), up to 60 cm high, is found on roadsides, walls, etc. in Europe, W. Asia, North America, and N. Africa. Its name results from its abundance after the Great Fire of London in 1666. the name is also given to plants of other related genera. Family: *Cruciferae*.

Rockets. Vehicles or missiles powered by jet propulsion that carry their own fuel and oxidizer and can therefore travel both in space and in the atmosphere. Rockets have no lift surfaces (see aeronautics), obtaining both lift and thrust from their propulsive jets. Solid-fuel firework rocekts were known to the Chinese in the 13th century and have been used sporadically in war ever since. However, the first liquid fuelled rocket to fly was designed by the American R.H. *Goddard* in 1926. This achievement attracted little interest, except in Germany, where the *Peenemunde* rocket research station was set up. It was here that Werner *von Braun* produced the V-2 rocket in World War II. Intercontinental *ballistic missiles* (with nuclear warheads), developed by both the USA and Soviet Union from this model, now form part of the armoury of all the principal nations. These rockets, in turn, led to the space rockets that carried man to the moon.

Space rockets are built in stages, a high-thrust vehicle providing lift-off and acceleration into the thinner atmosphere, where this first stage is jettisoned and the second stage takes over. The world's biggest space rocket, the US Saturn V, is a three-stage rocket weighing 3000 tonnes fully laden. The first stage burns kerosene and the second and third stages burn liquid hydrogen. Liquid oxygen (lox) is the oxidizer for all stages. This rocket can carry a load of 45 tonnes into space.

Fuel economy is one of the most important factors in rocket design—the specific impulse of a rocket being its most significant characteristic. The specific impulse is the time in seconds to burn one kilogram of fule while producing one newton of thrust. Typical values are 3400s for a solid fuel, 4500 s for liquid hydrogne-lox,and over 10 000 s for an ion engine. In an ion-engine rocket an easily ionized substance, such as caesium, is heated to produce ions, which are accelerated by an electric field to produce thrust. These engines of the future could be powered by nuclear energy or solar energy to produce the necessary 50 k W of electricity for each newton of thrust.

Roengten (R). Unit of ionizing radiation. One roengten of radiation induces 2.58×10^{-4} coulomb of charge per kg of dry air.

Rotameter. A device that measures the rate of flow of fluids.

Rotational Motion. Motion of a body that involves turning about an axis. If θ is the displacement, ω is the angular velocity, α the angular acceleration, the equation of motion for rotational motion are similar to the corresponding equation in linear motion, viz.

$$\omega f^2 - \omega t^2 = 2\alpha\theta, \ \theta = \omega_o t + \frac{1}{2}\alpha t^2 \text{ etc.}$$

The corresponding quantities for mass and force are moment of inertia and couple respectivley.

Rotor. Refers to the rotating part of an electric motor, dynamo or generator. *Also known* as armature.

S

Saccharimeter. An instrument used to measure the rotation of the plane of polarization optically active solutions.

Saturated Vapour. A vapour in equilibrium with its liquid at a given temperature. The presure exerted by a saturated vapour depends only on the temperature. For certain saturated vapours the specific heat capacity can be negative.

Scanning Tunnelling Microscope. A type of high-resolution microscope in which a small conducting probe is scanned slowly across the surface of a specimen. An electrical signal is produced as a result of electrons moving between the sample and the probe by the *tunnel effect.* The probe is raised or lowered to keep the signal constant, enabling a computer-generated contour map of the surface to be produced. The technique., which is used on conducting specimens, can resolve individual atoms.

Scattering. The spreading out or diffusion of a beam of radiation when it is incident on matter. The intensity of scattered light changes as

$$\frac{1}{\lambda^4}.$$

Screw. A simple machine when given a completed circular turn of radius *r,* it advances linearly by a distance *h,* known as its pitch. Due to friction, the efficiency of such a machine is low. The force ratio however can be quite high.

Search Coil. A small coil made up of several turns of wire that is used to measure strength of a magnetic field. The coil is suddenly withdrawn from the magnetic field, the current induced in the coil is proportional to the magnetic field strength.

Searle's Rod. (Searle's Apparatus). An apparatus for measuring the thermal conductivity of a conductor. A long rod of material is heated at one end by a temperature bath. The other end is cooled by a coil of copper through which a continuous steady stream of water is passed. The temperature of the rod is noted at two points along the length of the rod, separated by a distance l. If θ_1 and θ_2 are the temperatures at these two points and A is the area of cross section of the rod, the thermal conductivity of the rod is given by

$$\frac{kA(\theta_1 - \theta_2)}{l} = ms(\theta_4 - \theta_3),$$ when m is the mass of water flowing through the coil per unit time, s is the specific heat of water and θ_3, θ_4 are the temperatures of inlet and outlet of water.

Seconds Pendulum. A simple pendulum with time period of two seconds.

Secondary Colours. Colours produced by mixing pairs of primary colours.

Secondary Emission of Electrons. Electrons given out from a surface due to the impact of high energy electrons or ions. Such an emission takes place, in a photomultiplier tube.

Secondary Standard. Such a copy of a primary standard that the difference between the copy and the primary standard is known.

Secondary Winding. Refers to the winding on the load (i.e. output) side of a transformer irrespective of the transformer type.

Seed Crystal. A small crystal that can initiate the process of crystallization in a supersaturated solution or a supercooled liquid. It is used in the manufacture of semiconductor devices.

Self-Induction, Coefficient of (L). It is the magnetic flux linked with the coil when a unit current flows throught it. It is also equal to the e.m.f. induced in the coil when the current changes through the coil at the rate of 1 ampere sec. Its unit is henry.

Semi Conductor. A substance with conductivity, between that of an insulator and a conductor. The conductivity of such materials increase on increasing temperature as on a logarithmic scale. Pure semiconductors are known as *intrinsic semiconductors.* Their conductivity is usually small. The conductivity of semiconductors can be increased by addition of suitable impurities to them. Such semiconductors are known as *extrinsic semiconductors* and the phenomenon of adding impurities is known as *doping*. If doping a semiconductor increases the number of electrons or negative charge carriers, the semiconductor is known as *n*-type semiconductor, and the impurity is known as a donor. In *p*-type semiconductors, the conductivity is increased by increasing the number of positive charge carriers or holes. This is done by doping the substance with materials, known as acceptors, their valency is one less than that of the semiconductor atoms.

Semiconductor Diode. A *solid-star* electronic device with two electrodes. It consists of a single p-n junction (*see* semiconductor). When the p-region is at a more positive voltage than the n-region (forward bias), the current flow increases exponentially as the voltage rises. In reverse bias, very little current flows until a sufficiently hight reverse voltage has built up to cause breakdown; the current then increase sharply. The diode is, therefore, commonly used as rectifier. The *Zener diode* is designed to break down at a specific reverse bias voltage, above which the voltage across it remains effectively constant. The higher the doping levels on both sides of the juctions, the lower the breakdown voltage. It is used as a volatage regulator.

Semiconductor diodes, which have largely replaced thermionic diodes, are also used to generate microwaves by the *Gunn effect*, to detect light in *photocells,* and to emit light in low voltage displays. The latter, called *light-emitting diodes* (LEDs), are widely used in calculators and digital watches. They emit radiation (light) when holes and electrons combine, the colour of the light depending on the material of the crystal.

Semipermeable Membrane. A memberane that allows certain molecules in a fluid to pass through while stopping others.

Semitone. The smallest pitch interval between successive tones of the present western musical scales.

Sensation Level. A measure of the intensity I of a sound with reference to the minimum audible intensity I_o. If the sensation level L is measured in decibles, then $L = 10 \log (I/I_0)$, I_0 is generally taken as 2.5 picowatts per square meter.

Sensitivity. It is the response of a phyical device to a unit change in the input.

Sensitivity of a Balance. Deflection of the beam of the balance because of per unit difference in mass (in mg) in its two pans. So if a difference of *k mg* in load in the two pans produces a deflection Q of the beam the sensitivity of the balance is given by Q/*k*.

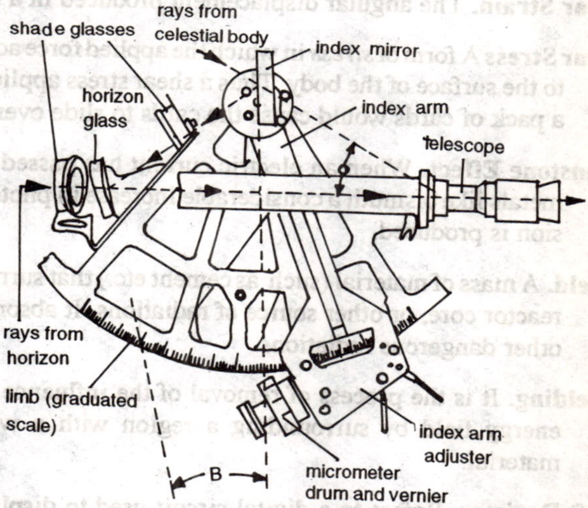

Sextant. Angle α measures the angle between horizon and reference arm; α/ β is the angle between index mirror and horizon glass, marked by the angular movement of the index arm along the limb α = 2β, therefore, the graduations marked on the scale are twice the actual angular movement.

Sextant. An instrument used primarily in navigation for determining latuitudes by measuring the angle subtended by some celestial body to the horizon. Thomas Godfrey of Philadelphia and John Hadley of London, working independently, discovered the sextant's principle in 1730. The graduated metal strip, shaped in an arc of the sixth part of a circle, gave the instrument its name. In use the movable index arm is slid along the scale until the image of the reference star as viewed in the half-silvered index mirror is aligned with the horizon. The reading on the scale then indicates the angle subtended.

Shadow. A region where light from a source is obstructed to reach by an opaque obstacle. The region of complete darkness is known as *umbra* and the region of partial darkness is known as *penumbra.*

Shear Strain. The angular displacement produced in a body.

Shear Stress A form of *stress* in which the applied force acts tangentially to the surface of the body. Thus a shear stress applied to the top of a pack of cards would cause the cards to slide over each other.

Shenstone Effect. When an electric current has passed through some metals like bismuth a considerable increase in photoelectric emission is produced.

Shield. A mass of material (such as cement etc.) that surround a nuclear reactor core, or other source of radiations. It absorbs neutrons or other dangerous radiations.

Shielding. It is the process of removal of the influence of an external energy field by surrounding a region with a wall of suitable material.

Shift Register. Refers to a digital circuit used to displace a set of an information in the form of pulsed either to the right or left.

Shock Wave. A narrow region of a high pressure in a fluid, created when a fast-moving body passes through the fluid. The waves are propagated outwards from the body and occur, for example, when an aircraft passes through the sound barrier.

Short Circuit. When the terminal of a cell are joined externally by a wire having negligible resistance, the cell is said to be short-circuited.

Shower. Refers to a group of elementary particles and photons arising from the impact of a single particle of a high energy.

Shunt. An element, generally a low resistance, connected across another component so as to reduce the current flowing through the second component. When connected across an ammeter, it can increase its range.

Sideband. Refers to range of frequencies lying on either side of a modulated carrier wave. The width of the sideband both above and below the modulated wave is equal to the highest modulating frequency.

Siemens (S). The *SI* unit of conductance equal to the conductance between two points on a conductor when a potential difference of one volt between the points causes a current of one ampere to flow. Named after Ernst Warner von *Siemens*.

Sievert (Sv). The SI unit of dose equivalent being that arising when the absorbed dose of *ionizing* radiation is 1 J/Kg, taking into account the dimensionless factors required to compensate for different types of radiation having different biological effects.

Sigma Pile. A neutron source and a moderator having no fissile material, used to analyse the properties of the moderator.

Sign Convention. The following sign convention is commonly adopted for measurement of distance in an optical system.

(i) The incident rays are considered as travelling from left to right.

(ii) All the distance measured in the direction of incident rays are taken as positive.

Signal. In an electronic circuit or system. The changing parameter by means of which information is conveyed through the system.

Signal Generator. Any circuit or device which produces an electrical parameter that is adjustable and controllable. Most generally, a voltage source with variable amplitude, frequency and waveform is used. The term is generally applied to sinewave generators; those generating pulse waveforms are known as pulse generators.

Signal-to-noise Ratio. Refers to the ratio of one parameter of a wanted signal to the same parameter of the noise at any point in an electronic circuit, device, or transmission system.

Simple Harmonic Motion. Any oscillation performed by a body about some reference point so that the restraining force is directly proportional to its displacement from that point. Examples of simple harmonic motion include a *pendulum* swinging through a small angle and a vibrating string. The maximum displacement of the system is known as the amplitude and the time taken for one complete oscillation, the period. This displacement, x, at a time t is simply represented by the equation $x = A \sin \omega t$, where A is the amplitude and ω the angular frequency, which is related to the period T by $\omega = 2\pi / T$. Thus a graph of x plotted against t has the shape of a sine wave, i.e., it is sinusoidal.

Simple Pendulum. A small spherical bob tied to the end of a long string whose other end is fixed. When displaced from mean position, its time period is $T = 2\pi \; \frac{1}{g'}$ where l is the length of string.

Simulator. A device that copies behaviour of an actual system but is made from components which are easier, cheaper, or more convenient to manufacture. Simulators are frequently used for solving complex problems.

Simultaneity. Two events that occur not only at the same time but also at the same place.

Sine Wave (Sinusoidal Wave). A waveform that resembles the plot of sine of an angle versus angle. The equation of such a wave has one variable proportional to the sine of another.

Siphon. A bent tube in which one arm is shorter than the other; used for transferring liquid from a higher to a lower level. It depends for its action on the principle that the pressure forcing the liquid up the tube at one end is more than the pressure acting on the liquid at the other end. The liquid continues to flow till the levels in the two vessels become equal.

SI Units (Systeme International d' Unites). An international system of units, based on the *m.k.s.* system, used for all scientific purposes. It has seven base units (metre. kilogram, second, ampere, kelvin, candela, and mole) and two supplementary units (radian and steradian). All physical quantities are expressed in these units or in derived units consisting of combinations of these units, 17 of which have special names and agreed symbols. Decimal multiples of all units are expressed by a set of prefixes. Where possible a prefix representing 10 raised to a power that is multiple of 3 is used.

Skip Distance. The minimum distance at which a sky wave can be received. This arises due to a minimum angle of incidence at the ionosphere below which a sky wave is not reflected. This minimum angle is a function of the frequency.

Sky Wave. A radio wave which reaches a point after reflection from the ionosphere and not directly from the transmitter.

Slide Wire Bridge. A thin uniform wire of eureka or manganin, generally one metre long, stretched over a wooden board along a metre scale. Its two ends are soldered to copper strips A and B. A third copper strip C is fixed on the board. It is based on the Wheatstone bridge principle.

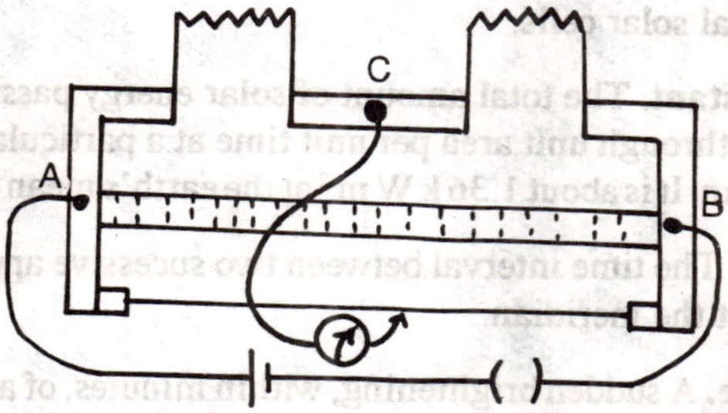

Sliding Friction. The force that slides—a body along a surface to overcome the force of friction.

Slow Neutron (Thermal Neutron). A neutron with kinetic energy less than 10 eV.

Slug. Unit of mass is the *f.p.s.* system. A mass that moves with an acceleration of 1 ft/s^2 when a force of 1 lb acts on it.

Snell's Law. When a ray of light passes from one medium to another the angle (r) between the refracted ray and a line normal to the interface between the media is related to the angle (i) of the incident ray, also taken to the normal, by the equation $\sin i/\sin r = n$, where n is the relative refractive index of the media (*see* refraction). It is named after the Dutch astronomer Willebrord Snell (1591-1626), who discovered it in 1621. It was not published until 1638, when Descartes announced it without crediting Snell.

Sodium Vapour Lamp. An inverted U shaped glass tube that is exhausted and filled with an inert gas (neon) at a pressure of about

10 mm. Specks of sodium metal are deposited inside in a scattered manner. Two tungsten spirals are scaled inside the two ends of the U-tube. On becoming hot, the tube emits yellow light of sodium.

Soft Iron. Iron containing a small amount of carbon. It is magnetized easily by placing it in a magneitc field but is demagnetize when the magnetic field is removed. Widely used in transformers, electric bells, electromagnets as core material.

Solar Cell. Any instrument that converts solar energy (light, heat, etc.) into electrical energy e.g., semiconductor devices operating by the photovoltaic effect and thermopiles. A solar *battery* is made up of several solar cells.

Solar Constant. The total amount of solar energy passing perpendicularly through unit area per unit time at a particular distance from the sun. It is about 1.36 k W m^{-2} at the earth's mean orbital distance.

Solar Day. The time interval between two sucessive appearances of the sun at the meridian.

Solar Flare. A sudden brightening, within minutes, of areas in the sun's atmosphere, resulting from an explosive release of energetic particles and radiation. Flares occur in regions of intense localized magnetic field, often above *sunspot* groups, and take up to an hour to fade. Large flares affect radio transmission on earth and produce *auroras.*

Solar Power. The use of the sun's energy to provide heating or to generate electicity. A vast amount of solar energy (about 3×10^{24} joules) falls on the earth every year. This energy can be converted into heat, the commonest method being by heating water flowing through special panels on the roof of a building. The temperature rise produced is small but it reduces the energy required from other sources for hot water and space heating. As some 45% of energy is used in space heating buildings in the UK, solar heating could make a considerable contribution to UK energy needs. Higher temperatures, to form steam for electricity generation, are possible using mirrors to focus the sun's rays. It is estimated that 7000 square meters of mirror are required to generate 1 megawatt by boiling water to drive a turbogenerator. Direct conversion of solar radiation into electrical energy is possible with *solar cells.* These consist of semiconductor juctions in silicon crystals that are sensitive to the **photovoltaic* effect. The method is used mainly in small-scale applications, for example powering remote monitoring equipment, spacecraft, marine beacons, etc. For this method to have widespread commercial applications the cost of solar cells would need to be reduced.

Solar Prominences. Immense clouds of gas in the solar atmosphere, denser and cooler than their surroundigs and visible, usually only spectroscopically, as flamelike projections beyond the sun's limb. They show great diversity in structure. *Quiescent prominences* persist for possibly several months at high solar latitudes, reaching, typically, a height and length of 40 000 km and 200 000 km. *Active prominences* are short lived in comparison and may alter their shape in minutes.

Solar System. A system comprising the sun and the astronomical bodies gravitationally bound to the sun, that is the nine major *planets,* their *satellites,* and the immense number of *minor* planets, *comets,* and *meteoroids*. Almost all the mass of the solar system (99.86%) resides in the sun. The planets orbit the sun in the same direction and, with the exception of Pluto, move in paths close to the earth's orbit (i.e. close to be *ecliptic*) and the sun's equator. This and other information is taken as evidence of the common origion of the sun and planets, some 4600 million years ago, following the contraction and subsequent flattening of a rotating cloud of interstellar gas and dust.

Solenoid. A coil of wire usually forming a long cylinder. When electric current flows through it a *magnetic* field is created. This field can be used to move an iron rod placed on its axis. Solenoids are often used to operate mechanical values attached to the iron rod by switching on or off the current. In most cars a soleniod is used to operate the heavy-current switch that supplies energy to the starter motor.

Solid Angle (Ω). The solid angle subtended by a surface at a point.

Solidifying Point. The temperature at which a liquid solidifies.

Solid State. One of the four states of matter and that to which all substances, except helium, revert at sufficiently low temperatures.

It is the only one in which matter retains its shape, as a result of the stronger intermolecular forces.

Solid-state Devices. Electronic devices made with solid *semiconductor* components. They have no moving parts and depend for their operation on the movement of charges within a crystalline solid. Solid-state devices are smaller, lighter, physically more robust, and more easily mass-produced than *thermionic* valve equipment.

Sonar (Sound Navigation and Ranging). A method of locating underwater objects by transmitting a high frequency sound pulse and detecting or receiving it after it is reflected from object. The method can be used to determine the depth of water at a location. The apparatus used is then known as echo sounder.

Sonometer. An instrument for studying the vibration of a fixed wire or string. It is made up of hollow wooden box having a wire stretched across its top. The wire is fixed at one end while the other end passes over a pulley and a load can be suspended from it. Any length of the wire can be set into vibration by placing two inverted V-shaped bridges at the ends, by placing vibrating tuning fork on the sonometer. Resonance is produced when the natural frequency of the vibrating wire.

$$\left(\text{as given by } f = \frac{1}{2l}\,\frac{T}{m}\right)$$

is equal to the frequency of the tuning fork. T is the tension in the wire and *m* is its mass per unit length.

Sound. A disturbance propagated through a medium by longitudinal waves. Strictly the term applies only to those waves that are audible to the human ear, i.e. with frequencies between about 20 and 20 000 hertz, those with frequencies above 20 000 hertz being called ultrasound and those below 20 hertz infrasound. Sound is propagated by vibrations of molecules in the medium, producing fronts of compression and rarefaction. Sound waves are longitudinal as the molecules vibrate in the direction of propagation; the speed of sound in air at 0°C is about 332 meters per second (760 mph). The three principal characteristics of a sound are its pitch (the frequency of the wave), loudness (the amplitude of the wave), and *timber* (the extent to which it contains harmonics of the fundamental frequency). However, there is a relationship between pitch and loudness.

Sound Barrier. An obstacle experienced by subsonic **aircraft* attempting to fly at or above the speed of sound. Drag increases sharply, lift falls off, and the aircraft becomes difficult to control. At subsonic speeds the pressure waves created by the aircrafts as it flies through the air are able to move ahead of the aircraft; at supersonic speeds they cannot escape in a foward direction as the source is moving faster than the pressure waves themselves. Thus shock waves build up on the aircraft's wings and fuselage, creating an apparent barrier to supersonic flight. The barrier was overcome for the first time by the US Bell X1 rocket aircraft in 1947. Since then many aircraft capable of supersonic flight have been built (including Concorde) by greater streamilining, sweptback wings, and more powerful engines. As these aircraft cross the sound barrier a *sonic boom* is heard. This is created by a shock-wave cone with the nose of the aircraft at its vertex. In level flight the intersection of this cone with the ground produces a hyperbola at all points along which the boom is heard.

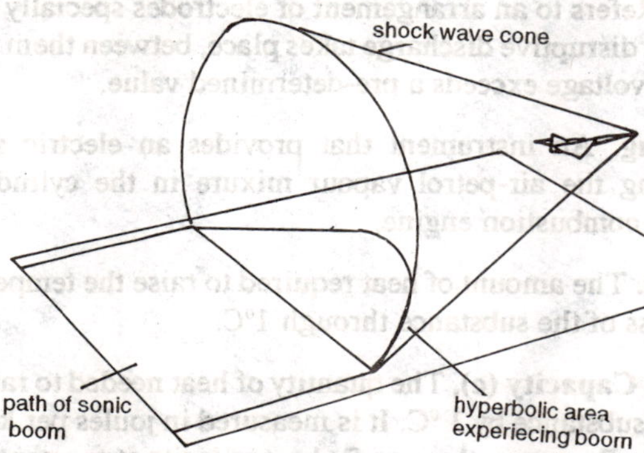

Sound Barrier. As an aircraft passes through the sound barrier the sonic boom is heard along a hyperbolic area on the ground.

Sound Intensity. The rate at which sound energy is propagated through a unit area perpendicular to the direction of propagation. It is measured in watts per square metre. The intensities of two sound levels are compared by a unit called the *decibel*. The intensity of sound is not the same as its loudness, the latter being the magnitude of the sensation produced by the human ear, which is dependent on the frequency of the sound.

Sound, Velocity of. The velocity with which sound waves travel in a medium. In air, its value is 332 *m/s* at 0°C and increases with temperature. It is independent of pressure. Its value is more in liquids and maximum in solids.

Space Charge. A region in a vacuum tube or semiconductor having some net electric charge on account of exess of dificiency of electrons.

Space Group. A set of operations (rotation about an axis, that when carried out on a periodic arrangement of points in space brings the system of points to self-coincidence.

Spallation. A vigorous unclear decay due to bombardment of a target by high energy particles, resulting in the emission of a number of alpha-and beta-particles.

Spark. A visible disruptive discharge of electricity between to places at opposite high potential.

Spark Gap. Refers to an arrangement of electrodes specially designed so that a disruptive discharge takes place, between them when the applied voltage exceeds a pre-determined value.

Sparking Plug. An instrument that provides an electric spark for exploding the air-petrol vapour mixure in the cylinder of an internal combustion engine.

Specific Heat. The amount of heat required to raise the temperature of unit mass of the substance through 1°C.

Specific Heat Capacity (c). The quantity of heat needed to raise a unit mass of substance by 1 °C. It is measured in joules per kelvin per kilogram. For gases, the specific heat capacity at constant pressure (c_p) exceeds that at constant volume (c_v) as heat is required to do work against the surroundings during the expansion. The ratio c_p/c_v (symbol: λ) is 1.66 for monatomic gases, 1.4 for diatomic gases, and about 1 for other gases.

Specific Heat of Gases. Unlike solids and liquids, the gases possess two specific heats one at constant pressure and the other at constant volume corresponding to the two ways in which gases can be heated. C_p is greater than C_v and $C_p—C_v=R$ where R is the gas constant and C_p, C_v represent the molar specific heats at constant pressur and constant volume respectively.

Specific Impulse. Refers to the ratio of the thrust produced to the rate of fuel consumption in a rocket.

Specific Inductive Capacity. Dielectric constant of a substance.

Specific Rotation. It is equal to the rotation of the plane polarisation which is per decimeter length of the solution per unit concentration of the solution.

Specific Thermal Capacity (Specific Heat.). Defined as the amount of heat which is required to increase the temperature of unit mass of substance through 1^0k. In SI units, it is measured in J/k g-k°. For a gas there are two specific heats. The reason is that heated by keeping either the pressure constant (when it expands) or the volume constant.

Spectral Series. Refers to a group of spectral lines in the spectrum of a substance. Such a series arises if electrons jumps from higher energy levels to a particular level e.g., for hydrogen atom, Balmer series results if electrons fall to the level $n = 2$ from higher level.

Spectrograph. An instrument wherein a photograph of a spectrum can be obtained.

Spectrometer. An instrument that is used for analysing the spectrum of a source of light. It consists of a *collimator,* that produces a parallel beam of light, telescope, that receives this beam after it passes through an analysing device like a prism, a grating etc., and a rotation table on which the device can be put. The position of the telescope can be read on a circular scale.

Spectrum. A set of spectral lines each corresponding to a different wave length. These are obtained when a beam of light is passed through a prism, grating etc. Such a set of lines may also be obtained when a susbstance is heated, or supplied energy by some other means. The electons in the atoms get excited to higher energy levels and emit radiations when they return to the original state.

Speed. The distance moved by a moving body in unit time. It is measured in *m*/sec and is a scalar quantity.

Spherical Aberration. A defect of image due to the paraxial and marginal rays which are coming to focus at different points on the axis of the lens. It can be corrected by using parabolic surfaces as reflectors and refractors.

Spherometer. An instruments which is used for the measurement of small thicknesses or the curvature of a spherical surface.

Spike. A current or voltage of extremely short duration.

Spin. A quantum number assigned to electrons and other elementary particles. It can have only two values either + 1/2 or —1/2.

Spontaneous Fission. Nuclear fission that occurs independently of external circumstances and is not initiated by the impact of a neutron, an energetic particle, or a photon.

Spreading Resistance. The component of resistance on account of the bulk of the semiconductor away from the junctions and contacts.

Spring Balance. Any instrument with which a force is measured by the extension produced in a helical spring. It is used in weighing. The extension produced is directly proportional to the force (weight).

Sputtering. Refers to the evaporation of particles of a cathode during the discharge of electricity through a gas. It can be used to coat a non-conductor with a thin metallic film.

Square Wave. A pulse train made up of rectangular pulses with mark-space ratio equal to unity.

Stability of a Balance. The tendencies of an oscillating balance to come to rest. The stability of a balance is opposed to its sensitiveness i.e. if stability of a balance is increased, its sensitivity decreases.

Stable Equlibrium. A body is said to be in stable equilibrium when it tends to return to its original state on being slightly disturbed from its state.

Standard Cell. A cell whose e.m.f. remains constant over a large length of time. It changes very little with change in temperature.

Standard Pressure. Pressure exerted by a column of mercury 760 *mm* high.

Standard Temperature. 273.15, 12, 0°C. Melting point of water.

Standing Wave (Stationary Wave). A wave pattern that results when similar wave motions travelling in opposite directions interfere. The pattern is characterized by the presence of modes, the point where the net displacement is always zero and antinodes, where the displacement is maximum. The distance between successive nodes, or antinodes is $\frac{\lambda}{2}$, λ being the wave length.

Stanhope Lens. A thick biconvex lens having front surface of radius two-thirds the thickness and a concentric back surface with radius one-third the thicknes.

Stanton Number. The reciprocal of the Prandtl number.

Star. A heavenly body that is huge source of energy. Energy production in stars is due to thermonuclear reactions.

States of Matter. Any of three distinct physical states: solid, liquid, and gas. A fourth state, *plasma,* is often added. The different sates are distinguished by the strength of the intermolecular forces compared to their random thermal motion. For example, in a solid the intermolecular forces are sufficiently strong to hold the molecules to an approximately fixed position, whereas in a gas the intermolecular forces have hardly any effect on the random movments of the atoms and molecules. Liquids represent an intermediate state.

Stat. A prefix which, when attached to the name of a practical electrical unit, means the corresponding unit in the CGS-electrostatic system.

Static Electricity. The effects created by electrical charges at rest. Current electricity is an effect resulting from a flow of electrons; in static electricity electrons from one object are pulled onto another object, usually by rubbing them together, but they do not flow. The effect can be observed with many nonconducting materials, for example a comb passing through dry hair or a leather sole on a nylon carpet. A force exists between two charged bodies (*see* electric field) attractive if they have opposite charges, repulsive if the charges are similar; the magnitude of the force is given by

Comlomb's law. Static electricity can ofter cause problems but can be useful in electrostatic precipitators.

Static Pressure. Pressure on a surface on account of a fluid at rest or due to another surface.

Statics. Branch of mechanics that deals with the behaviour of bodies under the action of forces, the bodies involved being at rest.

Steam. Water in the gaseous state. Water on being heated to its boiling poin (100^{0}C) is converted in to water vapour, known as steam.

Steam Calorimeter. A calorimerer wherein the amount of heat supplied is calculated from the mass of steam condensed on the body under test.

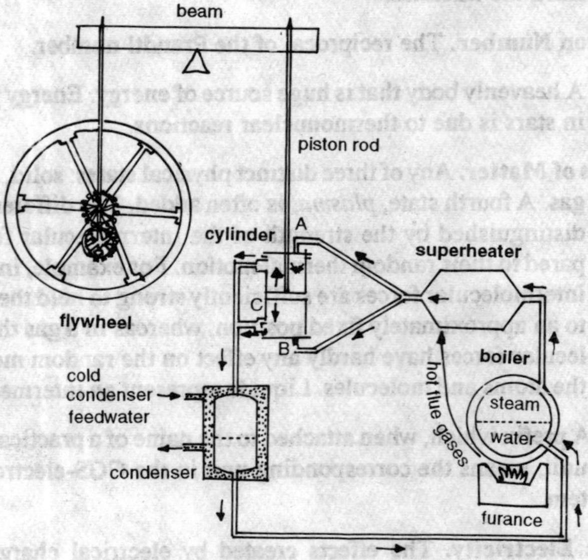

Steam Engine. The principle of a double-action beam engine of the kind patented by James Watt. During the first half of the cycle, valve A opens, steam flows in and pushes the cylinder down, and steam flows out to the condenser through valve C. During the second half of the cycle, B opens and the steam pushes the piston the other way.

Steam Engine. A heat engine in which heat from a furnace is used to raise steam, the expansion of which forces a piston to move up and down in a cylinder to provide mechanical energy. A primitive steam engine was invented in 1698 by a Capt Savery to pump water from mines. In 1711 *Newcomen* improved on this design but still relied on cooling the cylinder with a jet of water after each stroke. *Watt's* single-acting steam engine, patented in 1769, was the first to use a separate condenser. Watt went on to invent the double-acting engine, the crank and crosshead mechanism, and the governor. Thus by the end of the 18th century man had, for the first time, a reliable source of power that did not rely on muscles or the wind. It was largely this engine that created the *industrial* revolution. In 1808 Richard *Trevithick* made the first use of a steam engine to drive a carriage, but it was not until the end of the centry that steam cars were in use-and they were quickly replaced by cars using Otto's petrol engine. The steam engine, however, was the supreme prime mover of *railways* throughout the world from 1829 (when *Stephenson* built his first *locomotive*) until after World War II, when steam was largely replaced by electric and Diesel-electric trains. From the beginning of the 19th century steam engines were also widely used in place of sails in *ships*. Moreover, it was the steam engine that drove the first electricity generators for public supply. However, the more compact and efficient steam *turbine* has now replaced the steam engine for this purpose.

Steam Line. The curve showing the change of the boiling point of water with pressure.

Steam Point. Boiling point of water at one atmosphere pressure (100°C).

Stefan's Law. The energy emitted per unit area, E, of a black body per second is proportional to the fourth power of the absolute temperature T of the body. Thus $E= \alpha T^4$ where α is a constant known as Stefan's constant having a value.

$$5.73 + 10^{-8} Jm^{-2} S^{-1}\, {}^0K^{-4}$$

Step-down Transformer. A transformer that is used for decreasing an available voltage.

Setp-up Transformer. A transform that is used for increasing an available a. c. voltage.

Steradian (sr). The *SI* unit of solid angle equal to a solid angle that encloses a surface on a sphere equal to the square of its radius.

Stereoscope. An instrument that shows different views of the same scene to the two eyes giving the impression of depth.

Stimulated Emission. Principle on which the laser working is based. Transition of atomic electrons from higher to lower energy states on impact by photons of appropriate energy with the emission of radiation. Such an emission is known as stimulated emission.

Stokes. Symbol: St. The CGS unit of kinematic viscosity. 1 St = 10^{-4} metre squared per second.

Stokes (st). Unit of viscosity in the c.g.s. system.

Stokes' Law. The resisting force acting on a sphere, radius *r*, moving through a fluid under gravity with velocity *v* is $6\pi r\eta v$, where η is the *viscosity* of the fluid. The law is used nthe determination of viscosity. Named after Sir George *Stokes.*

STP (NTP). Standard temperature and pressure 0^0C, 1 atom.

Strain. In physics, the deformation of a body when it is subjected to a *stress. Longitudinal strain* is the extension per unit length when a body is stretched; *bulk strain* is the volume change per unit volume when a body is compressed; and *shear strain* is an angular measure of deformation.

Strangeness. A property of matter, expressed as a *quantum* number (s), postulated to account for the unusually long lifetime of some *hadrons.* In the quark model strange hadrons contain the strange quark or its antiquark. Strangeness is conserved in *strong* interactions and *electromagnetic* interactions.

Stratosphere. The region of the atmosphere above troposphere . In this region, the temperature remains nearly constant and the atmosphere is calm and undisturbed.

Streamlining. The process of shaping the contours of a body so that it presents the least resisntace to motion through a fluid. The mathematical study of these contours is part of *aerodynamics* and *hydrodynamics.* Streamlining must take into account drag forces, which are always present in real fluids, and is important in the design of aircraft, especially for supersonic flight.

Sress (Physics). The force per unit area that causea deformation (or *strain*) in a body. *Tensile stress* tends to stretch a body; *bulk stress* tends to compress it; and *shear stress,* which acts tangentially, tends to twist it.

Stroboscope. An instrument used to study periodic motion using a flashing light of known frequency. For example, if a disc revolving at 50 revolutions per second is illuminated by 50 hertz alternating-current mains lighting, it will appear to be stationary. This is because the eye, for any particular flash, sees the disc in exactly the same position as it was for the previous flash. Strobe lighting is used in discotheques to accentuate the music's beat, with which it flashes in time.

Strong Interaction. One of the four basic forces in the unitverse; it occurs between the class of elementary particles called hadrons. It is the strongest of the four (being 100 times stronger than the electromotive interaction) but is effective only over a very short range of about 10^{-15} metre. It is the force that holds the protons and neutrons together in the nucleus and is thought to occur as the result of the exchange of gluons between quarks.

Sub Carrier. A carrier wave used to under take another carrier wave.

Subcritical. A nuclear reactor in which the nuclear chain reaction is not self-sustaining.

Sublimation. Change of state from solid to vapours directly without changing to liquid.

Sub-shell. In Bohr's atomic theory, each electron shell is considered to be made up of various sub-shell. Each of these has the same orbital quantum number. The shells are known as *s*-shell, *p*-shell, ... etc.

Subsonic. Objects having speed less than the speed of sound i.e., for which the speed is less than Mach. 1.

Sun. A heavenly body that provides us with heat and light. Its mean distance from the earth is nearly 93 million miles. It is an approximately spherical bdoy having mass nearly 2×10^{27} tons and radius nearly 432050 miles. Its surface temperature is nearly 6000°C. While its interior temperature is nearly 13 million°C. The energy production in stars is due to thermonuclear reactions involving conversion of hydrogen into helium. The sun is a star.

Sunspots. Dark patches observed on the sun's surface that are regions of cool gas. Their presence is connected with local changes in the sun's magnetic field. They appear in cycles having a period of about 11 years.

Super Conductivity. The complete disappearance of electrical resistance when some substances are cooled to temperatures close to absolute zero. The transition to this state of zero resistance is sudden in many cases. The temperature at which this occurs is known as transition temperature The phenomenon can be used to produced large magnetic fields.

Super Cooling. Refers to the cooling of a liquid below its freezing point at the pressure. The effect can be obtained with pure liquids. Solidification occurs when a small impurity is added to the super cooled liquid.

Super Critical. A nuclear reactor having a self-sustained nuclear chain reaction.

Superfluid. A fluid that exhibits a very high thermal conductivity and virtually no friction at temperatures close to absolute zero. Such a fluid will flow up the sides and out of an open container. Liquid helium becomes a superfluid at 2.19 kelvins, called the lambda point.

Superheated Steam. Dry steam that has been heated to a temperature above 100°C. It is widely used as a working medium for converting heat into mechanical energy, for which purpose it is more efficient than normal steam.

Superheaterodyne. A system widely used in radio receivers; the incoming radio-frequency signal is combined with a locally generated carrier wave to give an intermediate frequency between radio-and audio-frequency (*see* modulation). The intermediate frequency is easier to amplify than radio-frequency, giving less noisy reception and better tuning.

Super Position Principle. Principle of super position.

Super High Frequency. Refers to frequency in the range.

$$3 \times 10^9 — 3 \times 10^{10} \text{ Hz.}$$

Super Sonic. Speeds larger than Mach 1 or more than the speed of sound.

Supplementry Units. Radian (Rad) and steradian (Sr.)

Surface Charge Density. Charge density.

Surface Colour. Coloured light reflected by a surface as distinct from the more common body colour which arises from reflection after some penetration into the medium. Transmitted light by bodies showing surface colour is complementary to the reflected colour.

Surface Energy. The energy per unit area of exposed surface.

Surface Tension. A force occurring on the surface of a liquid that makes it behave as if the surface has an elasitc skin. It is caused by forces between the molecules of the liquid: only those at the surface experience forces from below, whereas those in the interior are acted on by intermolecular forces from all sides. Surface tension causes a meniscus to form, liquids to rise up capillary tubes, paper to absorb water, and dropletes and bubbles to form. It is defined as the force acting tangentially to the surface on one side of a line of unit length (newtons per metre) or as the work required to produce unit increase in surface area (joules per square metre).

Susceptibility (x). The ratio $\frac{M}{H}$ for a magnetic subtance where H is the applied magnetic intensity and M the magnetization produced in the substance. $x < 0$ for diamagnetic substances and $x > 0$ for paramagnetic and paramagnetic substance. Its very is high for ferromagnetic susbstance.

Synchrocyclotron. A type of *cyclotron* in which the frequency of the accelerating electric field can be varied to compensate for the relativistic increase in the mass of the accelerated particles. This enables energies of up to 500 MeV to be obtained.

Synchronous Motor. An A. C. motor that has a rotation speed proportional to the frequency of the power supply.

Synchrotron. A particle *accelerator*, similar to the cyclotron, in which protons or electrons are accelerated in a circular path by an alternating electric field. The frequency of the field is synchronized with the energy of the particles to counteract their relativistic increase in the mass. Proton energies of several hundred GeV have been attained in these devices.

Synchrotron Radiation. Electromagnetic radiation emitted in certain directions by a charged particele when the presence of a magnetic filed confines its motion to a circle. The particle has to be moving at speeds comparable to that of light for a noticeable amount of radiation to be emitted. Therefore, a high magnetic field is needed. Such fields are used in *synchrotrons*, a type of particle *accelerator*. The emission of radio-frequency radiation also occurs from interstellar gas clouds in radio galaxies and, by analogy, is also known as synchrotron radiation.

Synthesizer. A device that can reproduce the sounds of conventional instruments electronically or produce a variety of artificial tones. Electronic oscillators produce a range of signals, which after amplification and appropriate filtering are converted to sound waves, of which some have the characteristic resonances of musical instruments, some are pure tones, and some are arbitrary combinations of sounds. Individual circuits can be plugged in and out by the player, enabling a wide range of sounds to be produced. The Moog synthesizer, invented by robert Moog (1934-) in 1965, can play one note at a time and is controlled by a keyboard. More recent *polyphonic synthesizers* cab be programmed to produce any number of different tones simultaneously.

T

Tachometer. An instrument for measuring the speed of rotation of a shaft, such as revolution counter in a car. Centrifugal tachometers measure the force experienced by rotating masses. Others are electrical or magentic, measuring electrical current or force generated by small generator in the instrument.

Tangent Galvanometer. A galvanometer based on the tangent law. It is made up of a coil of wire of many turns (n) held in a vartical plane parallel to Eath's magnetic field H. A small magnetic needle pivoted at the centre of the coil is free to rotate in a horizontal plane. A current i flowing through the coil rotates the needle through an angle θ such that

$$i = \frac{Hr \tan\theta}{2\pi n}$$

Tangent Law. When a magnetic needle is simultaneously acted upon by two uniform magentic fields acting at right angles to each other, then the tangent of the angle of deflection of needle is equal to the ratio of the deflecting field to the restoring field.

Tape Recorder. A device for recording and playing back sound stored on magnetic tape. In recording, the sound is converted to an electrical signal by a *microphone* and then amplified before being fed to an electromagnet in the recording head. The varying field of the magnet leaves a pattern of magnetization in the iron (or sometimes chromium) oxide coating of the tape as it passes through the machine. To play back, the magnetized tape induces a current in a coil as it passes the reproducting head. The coil current is then amplified and fed to loudspeakers. Tape recorders provide a compact and portable means of recording sounds of all kinds; the tapes are usually wound into *cassettes.*

Telecommunications. The transfer of information by any electromagnetic means, such as wire or radio waves. It includes telephones, telegraphy, *radio, televison,* etc. Generally, a telecommunications system consists of a transmitter, a transmission channel, and a receiver. The input to the transmitter is usually coded in some way and then fed to a modulator, in which it is combined with a carrier signal.

The transmission channel may be a wire, an optical fibre, or radio waves within a specified frequency range (the bandwidth). A single channel may carry several signals if a *multiplexer* is used. The receiver demodulates the signal and decodes it, converting it into the desired form of output, which may be sound, an electrical signal to a computer, a teleprinter printout, etc. Distortion may occur in the transmission because the signal has taken two different paths (shadowing) or because the channel may not respond equally to all the frequencies being transmitted. Noise (spurious signals) may arise at any stage of the transmission and reception process.

Telegraph. Transmission of measages using electrical impulses sent through wires. Messages are transmitted using a universally accepted code (called Morse Code) of dots and dashes that are identified by the length of time for which the current passes through a circuit.

Telemeter. An instrument that is used for recording of physical events at a distance.

Telephone. Which was invented by Alexander Graham Bell in 1875, carries speech in the form of electrical signals along a wire. A carbon *microphone* in the mouthpiece produces an electrical signal that passes through a network of exchanges and relays to the receiving earpiece, where it is converted back into sound by a small diaphragm *loudspeaker*. Telephone connections are now normally made automatically, first going through a local exchange and then, for long-distance calls, through the trunk network. All the switching operations are activated by the caller and controlled and monitored electronically. Telephone connectons exist to almost all parts of the world through undersea cables and *communications* satellites. *Telegraphy* is the transmission of written or printed message by electrical signals and was developed before the telephone, in 1837. The *Morse code* was used in the first commercial recordings telegraph, between England and France in 1851. Now telegrams, *Telex,* and *fax* messages are carried along the same wires as telephone conversations, using different frequency ranges. Radio telegraphy uses radio waves instead of a wire to carry the message. Like long-distance telephone links it may be relayed by communications satellites. Optical fibres transmitting beams of laser light are now in use and can carry considerably more information than an electric cable of the same thickness.

Telephoto Lens. A lens that is used for seeing magnified image. It is a combination of a convex and a concave lens and replace the lens ordinarily used in the camera to obtain the magnified image. The combination in increases the effective focal length of the lens.

Telescope. An optical instrument that produces a magnified image of distant objects. It is used on land and a major research tool in astronomy. The first *reflecting telescope* was produced by *Newton* in 1668. In the reflecting telescope the light from an object is collected by a concave, usually paraboloid, mirror of long focal length. This primary mirror reflects the light into a secondary optical system, which in turn reflects it in to a short-focus eyepiece. The eyepiece lenses produce a magnified image that can be viewed by eye, photographed, or otherwise analysed. Depending on the secondary optics, reflectors are called *Gregorian, Newtonian, Cassegrain,* or *coude telescopes.*

The *refracting telescope* was invented in 1608 by *Lippershey* in Holland and developed by *Galileo* as an astronomical instrument a year later. The light in the refracting telescope falls on a converging long-focus objective lens. The resulting image is then magnified by the short-focus eyepiece to produce the final image. An example is the *Keplerian telescope,* which was the first major improvement on Galileo's original design. Refractors are also used as terrestrial telescopes, usually containing an additional lens or a prism to cause the inverted image to be seen erect.

A telescope's light-gathering power depends on the area of the primary mirror or the objective lens. Since large mirrors are easier to fashion and mount than large lenses, the major astronomical telescopes are reflectors. In addition, unlike the objective in refractors, the reflector's mirror suffers no chromiatic *aberration,* and its spharical aberration and coma are minimal. The Hubble space telescope, launched in 1990, is 350 times more powerful than other earth-based telescopes.

Teletext. An information service in which pages of text are transmitted together with normal television broadcasts for display on a modified domestic television set. The system utilizes two of the unused lines between picture frames. All the stored information, which may be weather reports, news flashes, sports results, etc., is sent out continuously in a cycle at a rate of 50 lines pe second (2 per picture frame). The required pages are selected for display by keyboard, and the user may have to wait a few minutes for them to come up on the screen if the cycle is a long one containing a large number of transmission. The two systems in use in the UK are *Ceefax,* run by the BBC, and *Oracle,* run by the Independent Broadcasting Authority.

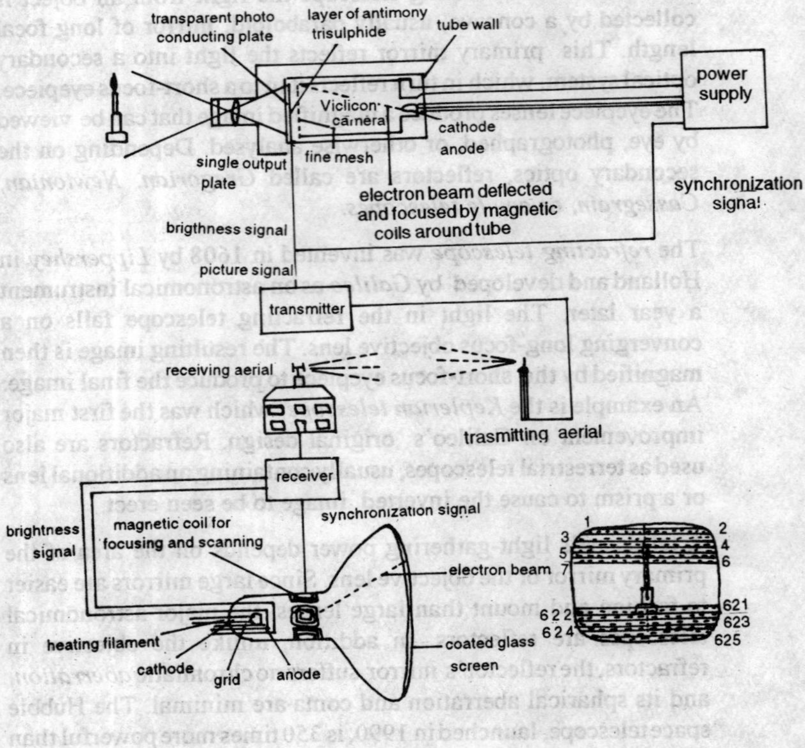

Television (TV) The broadcasting of pictures and sound by *radio* waves or electric cable. Television was invented by John Logie *Baird* in 1926. At a TV studio or outside-broadcast unit, a television camera converts the picture into an electrical signal. In most of Europe the picture consists of 625 lines made by an electron beam scanning the screen of a *cathode-ray* tube, 25 such pictures being formed every second. In the USA, 525 lines and 30 frames per second are used. A brightness signal and a synchronization signal (to form the lines and frames of the picture) make up the picture signal, which is used to modulate (*see* modulation) a VHF or UHF carrier wave and is broadcast with the modulated sound carrier wave (which has a slightly different frequency). In colour TV, a colour signal has to be added to the picture signal.

The aerial of the TV receiver detects the broadcast radio waves and the picture and sound signals are separated within the receiver. The picture signal is demodulated, and the resulting current is used to control the electron beam in a cathode-ray tube so that the picture is reconstructed, the scanning being sufficiently fast to give the impression of a continuous moving picture. A dish aerial is used for receiving satellite broadcasts.

In a *colour television* camera light from the scane to be televised is filtered into three primary colour components; Red. green, and blue. Light of each colour goes to a separate image tube.

There are three systems in use for encoding colour picture information for transmission: the US system, also used in Canada, Mexico, and Japan; the PAL (phase alteration line) system used in the UK and most other W European countries; and SECAM (systeme electronique couleur avec memoire), used in France, E Europe, and the Soviet Union. All of these combine colour and intensity information with sound and synchronization in a similar way to black-and -white television. The colour television receiver splits the signal into red, green, and blue components and applies these to three separate electron guns. The beam from each gun activates a set of phosphor dots of that colour on the screen, thus reconstructing the red, green, and blue components of the picture.

Telex. A telegraphy system using telephone lines to transmit printed messages from one terminal (the teleprinter) to another. The message is typed onto a keyboard transmitter, which converts the characters into a coded electrical signal for transmission. A printing receiver (combined with a the transmitter) carries out the reverse process, typing out messages as they arrive. Telex messages are directed through the telephone lines by the subscribers' numbers, each subscriber also being identified by his call sign. The system is extensively used by commercial and other organizations, because it costs less than a telephone call and provides both parties with a written copy of the messages. To increase the speed of transmission most teleprint are equipped with a paper-tape punch to prerecord messages.

Temperature. A physical quantity that is a measure of the average *kinetic* energy of the constituent particles of a body. It determines the direction in which *heat* flows when two bodies are in contact, the body with the higher temperature losing heat to that with the lower. The temperature of a body is measured in kelvins, degrees Celsius, or degrees Fahrenheit either by a *thermometer,* if the temperature is below about 500°C, or, if above, by a *pyrometer*.

Temperature Coefficient. A constant that determines how a physical quantity changes with temperature e.g. the temperature coefficient of linear expansion tells by how much the length of a substance change, due to change of 1^0 in the temperature.

Temperature of Inversion. (T_1). The temperature of the (hot) junction in a thermocouple at which the thermo e.m.f. changes sign. It has been found to depend on the temperature of the cold junction of an object of a given material and change with temperature.

Temperature Scale. A scale of temperatrure used for measuring temperatures. It has two fixed points, corresponding to two easily reproducible systems that are assigned two fixed temperatures, the interval between them is divided into an equal number of parts. e.g., on the celsius scale, the fixed temperature are the ice point (called 0°C) and the steam point (called 100°C) and the interval between them is divided into 100 equal parts, each interval is 1°C.

Temporary Magnetism. Magnetism induced by a magnetic field that dispapears as soon as the magnetic field is removed.

Tensile Strength. The ability of a material to withstand a "pulling" stress (*tensile stress*). It is usually measured by taking a bar of the material and stretching it to destruction. The ultimate tensile strength is the maximum load divided by the bar's cross-sectional area and is measured in newtons per square metre.

Tera (T). A prefix meaning 10^{12}.

Terminal. A point of an electrical circuit where connecting lead can be connected.

Terminal Speed. The steady reached by a body when it falls freely through a viscous fluid.

Terminal Velocity The maximum velocity attained by a body falling through a fluid. It occurs when the drag balances the gravitational force on the body: until the terminal velocity is reached the falling body accelerates; thereafter the velocity remains constant. According to Stokes' law the terminal velocity of a sphere (radius r) falling through a fluid with a coefficient of viscosity η is $2\phi r^2 g/9\eta$, where θ is the difference between the body and fluid densities and g is the acceleration of free fall.

Terrestial Magnetism. Earth's magnetism.

Terrestial Telescope. Telescope.

Tesla. (T) SI unit of magnetic flux density.

$$1\ T = 1\ Wb/m^2.$$

Tetrode. A thermionic valve that contains four electrodes. It has a screen grid between the control grid and anode of a triode valve. This improves the performance on a triod valve as an amplifier and oscillator at high frequencies.

Theodolite. A telescope fitted with spirit levels and angular scales used to measure altitude and azimuth; it is used in surveying, etc.

Theorem of Parallel Axis. If I_0 represents the moment of inertia of an object about an axis, the moment of inertia about an axis parallel to the given axis and at a distance *d* from it is given by $I = I_0 + md^2$ where *m* represents the mass of the object.

Theorem of Perpendicular Axis. If I_x, I_y are the moment of inertia of a body about *x*- and *y*-axis respectively, the moment of inertia about an axis perpendicular to the *x-y* plane and passing through the point of intersection of *x*- and *y*-axis is equal to (I_f) given by

$$I_f = I_x + I_y.$$

Therm. A unit of heat (energy), equal to 10^5 B. Th. U. 1 Therm = 1.055 J.

Thermal Agitation. Refers to the random movement of the molecules of a substance.

Thermal Capacity (also called Heat Capacity). The amount of heat (energy) needed to increase the temperature of a body by 1^0 K. It is equal to the produce of the mass of the body and its specific heat. It is measured in joule/^{0}K.

Thermal Diffusion. A method of separating heavier gas molecules form lighter ones by maintaining a temperature gradient over a volume of a gas containing particles of different masses. It is also used for separating for the isotopes of an element.

Thermal Equilibrium. State of a system where in no heat flow takes place between its components. If two bodies are in thermal equilibrium, they have the same temperature.

Thermal Neutrons. Neutrons having slow speed and energy; such neutrons are of importance in nuclear fission reactions in nuclear reactors. They have energy 0.1 eV.

Thermal Reactor. A nuclear reactor (see nuclear energy) in which natural or enriched uranium is used with a moderator, in most cases to generate heat for a *power station*. The moderator slows down the neutrons emitted during the fission of uranium-235, so that their velocities are comparable to the thermal velocities of gas molecules. In a thermal reactor, fuel rods made of uranium metal or oxide are surrounded by a moderator (together forming the reactor core), the heat of the reaction being removed by a coolant. After leaving the core, the coolant passes to a heat exchanger in which steam is raised. The rate of reaction is controlled by control rods, moved in and out of the core: these contain a neutron-absorbing element, e.g. boron or cadmium. The first nuclear reactor was built by *Fermi* in 1942, using natural uranium a graphite moderator, and a water coolant. Early British thermal reactors were the Magnox type, using natural uranium; 18 were built, the earliest of which are now being decommissioned. The next generation of British reactors were the advanced gas-cooled reactors (AGR), using enriched uranium dioxide pellets, each about the size of a thimble and equivalent to 1.25 tons of coal. However, thermal reactors are now usually based on the US-designed pressurized-water reactors (PWR), using ordinary water as both modrator and coolant. These are now considered the most economical and efficient; they also power all nuclear submarines.

Thermionic Emission. The emission of electrons from the surface of substance on being heated. The number of electrons emitted from a hot surface per unit area depends on the temperature T of the surface and is given by $AT^2 \bar{e}\Phi/kT$ where A represents a constant Φ is the work function of the surface material.

Thermionic Valve. A device consisting of a sealed glass or metal tube, either evacuated or containing gas at low pressure, into which two or more electrodes are inserted. One heated electrode, the cathode, emits electrons, which are attracted to the positively charged anode, forming an electric current. This current flows in one direction only and can be controlled by the voltage applied at one or more other electrodes (called grids). The *diode valve* was invented in 1904 by Sir John Ambrose Fleming. It has two electrodes and functions as a simple *rectifier*. The *triode valve* with one grid was invented in 1910 by Lee De Forest; it was the first to function as an amplifier. The weak signal fed to the grid produces a stronger signal in the anode circuit. Diode and triode valves made possible the development of radio transmitters and receivers, although now *semiconductor* diodes and *transistors* have replaced them.

Thermionics. Study of emission of electons by substances on being heated.

Thermistor. A *semiconductor* device with an electrical resistance that decreases sharply as temperature increases. It is used in temperature-control circuits and as a thermometer.

Thermocouple. A device which is used for measurement of temperatures. It is made up of two metallic junctions of dissimilar metals. It is based on the fact that when the two junctions are placed at different temperatures, an e.m.f. develops across them, its magnitude depends on the temperature difference between the junctions. A measurement of the e.m.f. helps one to calculate the desired temperature.

Thermodynamics. The study of heat and it relationship with other foms of energy. Thermodynamics is primarily a statistical subject, thermodynamic quantities, such as *temperature* and *entropy,* being dependent on the statistical behaviour of the particles that comprise as system. There are three fundamental laws of thermodynamics. The first law states that the energy of a closed system remains constant during any process. This law is a restatement of the law of conservation of mass and energy. The sedond law states that heat ċannot flow from a cold body to a hot body without the expenditure of external work. Another way of stating this law is to say that the entropy of a closed system can never decrease. It the enetropy change is zero then the process is said to be reversible. The third law states that as the thermodynamic temperature of a system approaches *absolute zero,* its entropy approaches zero. Thermodynamics has proved most valuable in studying *heat* engines and chemical reactions, but it also has wider implications in other statistical sciences.

Thermodynamic Temperature (T). Thermodynamic temperature of triple point of water is taken at 273.16 K. Temperatures on the celsius scale (c) and kelvin scale (k) are related as C + 273 = K. The size of the temperature on the two scales is equal.

Thermoelectric Power. Refers to the rate of change of the thermo-e.m.f of a thermocouple with the change of temperature of the hot junction.

Thermometer. An instrument for measuring temperature. Thermometers used a property of a substance that varies uniformly with temperature, most commonly the expansion of liquid mercury. A clinical thermometer is a typical mercury-in-glass thermometer; it has constriction in the stem above the bulb so that the mercury stays at its maximum reading until "shaken down". More accurate thermometers use the expansion of a gas, which is much greater than that of a liquid. Other thermometers include the resistance thermometer, which depends on the variation in the resistance of a wire (usually platinum); the bimetallic strip in which the unequal expansion of two metals welded together causes a pointer to move round a dial; and the *thermistor*. High temperatures are measured by a pyrometer.

Thermometer Clinical. A mercury thermometer for measuring the temperature of the human body. Its range is normally from 95^{0}F to 110^{0}F. It is provided with a constriction in the tube near the bulb to prevent the mercury from contracting when the thermometer is taken out from the patient's mouth.

Thermometer Maximum and Minimum. A thermometer that records the highest and the lowest temperature reached during an interval of time. It is made up of glass bulb filled with alcohol, that on expansion, pushes a mercury thread along a graduated tube. A steel index at the end is pushed by the expanding mercury. It cannot fall back and so stays at the extreme position to which it has been pushed and is an indicator of the maximum temperature reached during the interval. The same principle has been used for recording the minimum temperature.

Thermo-pile. A device used for detection and measurement of heat radiation. It is made up of a number of thermocouples made of antimony and bismuth rods connected in series. When exposed to heat radiation, it produces a current that can be detected and measured.

Thermosphere. The region of the upper atmosphere where the temperature increases with height.

Thermostat. A device for maintaining a constant temperature, by cutting off the energy supply when the required temperature is reached and restoring it when the temperature falls. It usually consists of a bimetallic strip that bends when it expands and contracts, thus opening and closing an electrical circuit or a gas or liquid flow valve.

Thick Lens. A real lens, as distinct from a hypothetical lens. In a thick lens the separation between the two surfaces cannot be ignored.

Three Phase. Having three equal alternating voltages between which there are relative phase differences of 120^0.

Threshold Frequency. The minimum frequency that gives rise to the photoelectric effect. The value of the frequency depends on the metal exposed to the light beam.

Threshold of Hearing. That minimum intensity level of a sound wave which is audible. It occurs at a loudness of about 4 phons.

Threshold Voltage. The voltage where particular characteristic of an electronic device first occurs.

Thyristor. A solid-state electronic device, also called a semiconductor or silicon-controlled rectifier; it consists of four layers of *semiconductor* forming three p-n junctions. It acts as a switch, blocking the current through two terminals until it has been turned on by a pulse applied to the third terminal. This pulse can be initiated by light or a temperature change. Thyristors are used in a wide range of power-switching and control-circuit applications. They can pass currents ranging from milliamperes to several hundred amperes and for many purposes have replaced the thyratron.

Tides. Violent motions of the sea caused by the attraction exerted upon the sea by the moon, and to a smaller extend by the sun. At full and new moon the tidal force of the sun is added to that of the moon, resulting in high spring tides: while at half moons the forces are opposed, causing low-tides.

Timbre of Sound. The property of a note that enables us to assign it to tis source is known as the quality or timbre. It depends on the number and intensity of the overtones present.

Time. A concept that measures the duration of events and the periods that separate them. It is a fundamental parameter of all changes, measuring the rates at which they occur; it proovides a scale of measurement enabling events that have occurred to be distinguished from those that are occurring and those that will occur. It appears, intuitively, to be flowing at a constant rate in one direction only, for all observers. However, according to Einstein's theory of *relativity* this is not the case. The rate at which time passes (as measured by a clock) is not the same for observers in different frames of reference that are moving at a constant velocity with respect to each other. Thus, according to the *time-dilation* effect, if two observers are moving at a constant velocity relative to each other it appears to each that the other's time processes are slowed down. This means that events that appear to be sumultaneous to observers in the same frame of reference would not be simultaneous to observers in dirrerent frames of refernce. In order to pinpoint an event in the universe, its position in a four—dimensional space-time continuum must be specified. This continuum has three space dimensions and one time dimension.

Historically, the measurement of time on earth has been based on astronomical observations—the time taken for the earth to revolve on its axis (the day) or for it to complete its orbit round the sun (the year). However, in modern science the basis of time measurement is the *second*, which is defined in terms of the frequency of the radiation emitted in a specified transition of an isotope of caesium.

Time and Motion Study. A study of a machine and its operator in a factory, measuring the time taken to complete each action and the sequence of the operator's movements. The study is then used to recommend methods of speeding up the operation and reducing the effort involved in order to utilize labour as economically as possible.

Time Dilation. An effect predicted by Einstein's Special theory of *relativity*. If two observers, A and B, are moving at a velocity v relative to each other, it will appear to A that B's clock will show that time is running more slowly; thus a time t measured on A's clock will be $t(1-v^2/c^2)^{1/2}$ on B's clock, where c is the velocity of light. The effect has been observed in some particles moving at high velocities, which appear to have an anomalously long life-time.

Time Lag. *Syn.* time delay. The time that passes between the closing of one circuit in a circuit-breaker and the closing of the main current circuit.

Time Switch. A switch which incorporates a type of clock mechanism for making and/or breaking an electric circuit at times that are predetermined by the setting of the mechanism.

Tint. Colour.

Ton. An Imperial unit of weight equal to 2240 lb (long ton) or 1016 kilograms. In the USA a unit equal to 2000 lb (short ton) is also used. The metric ton (or *tonne*) is equal to 1000 kilograms.

Tone. (1) An audible note having no partials. (2) The quality of a musical sound. e.g. loud tone, soft tone.

Tone Control. Device for adjusting the frequency response of an audio-frequency amplifier so as to compensate for deficiencies in the audible output and therefore achieve a more pleasing result.

Tonne. Symbol: *t*. A metric unit of mass. 1 $t = 1000$ kg. Generally called a metric ton, it differs in size from an imperial ton (116 kg) by about 1-1/2%.

Toric Lens. A lens having a toroidal surface, i.e. a surface generated by a circular arc rotating about an axis which does not pass through the centre of the circle. The curvature in one plane is then different from that in an orthogonal plane. These lenses are used for correcting artigmatism in the eye.

Torpedo (*armament*). A self-propelled guided underwater missile carrying a high explosive warhead. They can be launched by ships or aircraft but have been used most successfully by submarines. Designed in 1866 by a British engineer, Robert Whitehead (1823-1905), they were used to sink 25 million tons of Allied shipping in World Wars I and II. Modern torpedos are driven by steam turbines or by battery-powered electric motors and have sophisticated active or passive acoustic homing systems (active devices send out sounds and are guided by the echo from the target; passive devices are guided by sounds from the target).

Torque (τ). The torque of a force $\vec{F}$ about an axis is equal to $\vec{r} \times \vec{F}$ where $\vec{r}$ is the radius vector from the axis to the point of application of the force. It is measured in N-m. Also known as the moment of force.

Torr. A pressure equal to the pressure of 1 mm Hg.

Torricelli's Theorem. The velocity of a fluid coming out of a vessel at a point distance *h* below, its surface, is given by

$$v = \sqrt{2gh}.$$

Torsion. The angular strain produced in a body on being subjected to a torque.

Total Internal Reflection. When light passes from a denser medium to a rarer, the ray is bent away from the normal. If the incident ray meets the surface at such an angle that the refracted ray must be

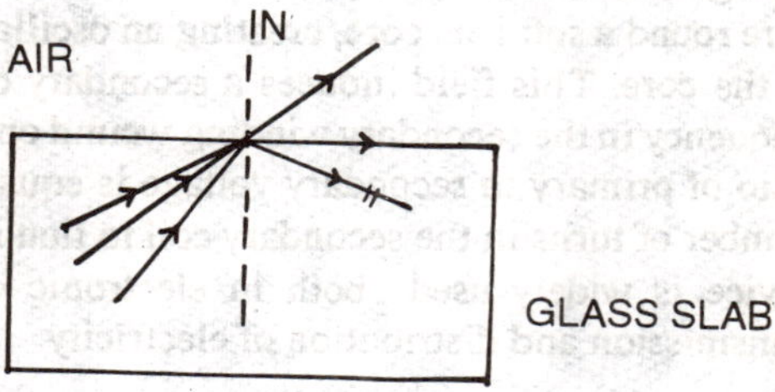

bent away at an angle larger than 90^0, the light cannot emerge at all and is said to have undergone total internal reflection.

Tourmaline. A mineral crystal which bits dichroic properties, i.e., doubly refracting with selective absorption of the ordinary ray while transmitting the extraordinary ray.

Tracer Bullet. A bullet that when ignited by the propellant emits light or smoke. In flight, its path appears as a continuous streak enabling the gunner to correct his aim. Mixed with other types of bullets during loading, tracer bullets are used in aircraft and by ground troods.

Track. Refers to an illuminated track of an ionizing particle created in a cloud chamber, bubble chamber, etc.

Tracking. (1) An arrangement by which an electrical parameter of one device varies in sympathy with the same or another parameter of the second device, when both devices are subjected to a common stimulus, (2) The formation of electrically conducting paths on the surface of solid dielectrics and insulators on being subjected to high electric field.

Transconductance. Refers to the mutual conductance between the control grid of a thermionic value and its anode; it is expressed in mhos.

Transducer. Any device that changes a signal or physical quantity into another form. *Microphones, loudspeakers* and *thermocouples* are examples. It is also usually the primary sensor in a measurement or sensing system.

Transformer. A device for converting alternating current from one voltage to another. The input is fed to a primary winding, a coil of wire round a soft iron core, creating an oscillating magnetic field in the core. This field induces a secondary current of the same frequency in the secondary winding wound on the same core. The ratio of primary to secondary voltage is equal to the ratio of the number of turns in the secondary coil to that in the primary. The device is widely used both in electronic circuits and in the transmission and distribution of electricity.

Transistor. A *semiconductor* device with three or more electrodes. Transistors form the basic elements of electronic *amplifiers* and logic circuits, often combined with other components in *integrated* circuits. They were first developed in 1948 by William *Shockley* and his coworkers at the Bell Telephone Co and now replace *thermionic* valves in most applications. The term transistor usually refers to the *bipolar junction transistor,* which consists of two junctions between p-type and n-type semiconductors forming either a p-n-p or n-p-n structure. Current is caried across these junctions by both negative and positive charge carriers (electrons and holes). The current between the emitter and the collector electrodes varies not only with the voltage drop across them, but also with the voltage or current level at the base (the third electrode). Depending on how it is connected into a circuit, the junction transistor can act as a voltage or current amplifier in much the same way as a triode valve. Transistors, however, work at a much lower voltage than valves, are more compact and robust, emit less heat, and are cheaper to make.

Originally, junction transistors ware made by alloying the impurity metal onto the semiconductor crystal or by adding impurities as the crystal was being grown. Now the doping is diffused in as a gas, introduced by ion implantation, or more commonly, achieved by a combined process of etching and diffusion, known as the planar process.

The *field effect transistor* (FET) is a unipolar device, in which current is carried by only one type of charge. There are two types: the junction FET (JFET) has a region of semiconductor of one doping type flanked by two highly doped layers of the opposite type. Current flows parallel to the junctions, between the so-called source and drain electrodes, through a narrow channel between the highly doped regions (the gate); it is controlled by the electric field arising from the gate input voltage, which alters the width of the conducting channel. The JFET is used as a separate compnent in amplifiers and switches. In the insulated-gate FET (IGFET) the source and drain electrodes are higly doped regions in a substrate of the opposite type. The gate electrode is a conductor separated from the substrate by a thin insulating layer across the surface. The electric field caused by the gate voltage controls the source-drain current on the other side of the insulator. The IGFET is used mainly in metal-oxide semiconductor (MOS) integrated circuits. It is smaller than the equivalent bipolar junction transistor and uses less power. The first plastic transitor was built in 1988.

Transit Instrument. A telescope theodolite, or similar device, that is used to measure angular position. In astronomy, for instance, transit telescopes measure the vertical angles of stars or planets.

Transition Temperature (Transition Point). Refers to the temperature at which one form of a substance changes into another; the temperature at which both forms could co-exist.

Transit Time. The time taken by an electron to pass directly from one specified point another under given operating conditions.

Translatory Motion. A motion which involves a non-reciprocating movement of matter from one place to another.

Translucent. Any susbstance that permits the passage of light in such a way that an object cannot be seen clearly through the susbstance.

Transmission Coefficient, Transmittance (T). If a beam of light is alowed to pass through a medium the radiation is absorbed to a greater or lesser extent so the intensity of the beam decreases. The ratio of the intensity after passing through unit distance of the medium to the original intensity is known as the transmission coefficient.

Transmission Density. *Syn*. optical density. Symbol: D. The logarithm of the reciprocal of the transmittance, i.e., $D = \log_{10} 1/\tau$.

Transmittancy. The ratio of the transmittance of the solution to the transmittance of the same path length of solvent.

Transmitter. The device, apparatus, or circuits by means of which the signal is transmitted to the receiving parts of the system in any telecommunication system.

Transmutation of Elements. The process of converting one chemical element into another. Aritificial transmutation can be done to some extent with the help of suitable nuclear reactions.

Transparent. Permitting the complete passage of light so that objects can be seen clearly through them.

Transponder. In telecommunications, a combined transmitter and receiver that sends out a signal automatically on receiving a predetermined trigger signal.

Transuranic Elements. Elements having atomic number greater that 92. All of them are redioactive.

Transverse Wave. A wave motion having the vibrations perpendicular (transverse) to the direction of propagation of the wave.

Tribology. The study of *friction* and such allied topics as lubrication, abrasives, surface wear, etc. It includes such effects of frictions as triboluminescence and frictional elecricity.

Triboluminescence. Luminescence that results form friction.

Trigger. Any stimulus which initiates a particular response from an electronic circuit or device.

Trimmer. *Syn*. Trimming capacitor. An adjustable capacitor with a small maximum capacitance in parallel with a fixed capacitor to enable the capacitance of the combination to be adjuted.

Triode. A thermionic value consisting or three electrodes a cathode, a grid and an anode (or plate). The cathode is heated and acts as a source of electrons; the anode acts as a collector of electrons by applying a high voltage to its and the grid, also known as control grid, serves to regulate the quantity of electronic current flowing through the triode by the application of a small voltage to it. The device can be used to amplify small signal voltage or to produce electromagnetic oscillations.

Triple Point. It is the temperature at which the gas, liquid and solid phases of a substance can co-exist in equilibrium The triple point of water is taken as 273.16 k.

Tritium. ($_1H^3$): An isotope of hydrogen having 1 proton and 2 neutrons in the nucleus. It is a radioactive substance having a half life of 12.3 years.

Tropopanse. The layer of atmosphere that separates Troposphere and Stratosphere.

Troposphere. Region of atmosphere lying just above the earth's surface. It extends in height to about 10 miles at the equator and about 5 miles at the poles. The temperature in this region decrease with height.

Trough . The maximum disdance below the line of mean position in a transverse wave. Two successive troughs are separated by a wave length of the wave.

Tuning Fork. A metallic fork having two prongs which, when struck, from a pure sound tone.

Tunnel Effect. The passage of an electron or other particle through a potential barrier when, according to classical menhanics, it has insufficient energy to do so. It is explained by *wave* mechanics on the basis that the electron is not completely localized in space, part of the energy of the associated wave being able to tunnel through the barrier. The effect has a negligible probability in large-scale systems, but a finite probability in microscopic systems. It is the basis of some radioactive decay processes and is made use of in the *tunnel diode,* a semiconductor device that has a negative resistance over part of its operating range.

Turbine. A device in which a moving fluid drives a wheel or motor, converting the kinetic energy of the fluid into mechanical energy. In its simplest form it is known as a *water wheel,* which has been in use since ancient times to drive mills, pumps, etc. The principle of the war wheel forms the basis of the hydraulic turbine, used in the generation of hydroelectricity. The most widely used types are the Pelton Wheel patented in 1889 by Lester Allen Pelton (1829-1918),. the radial Francis turbine, designed in 1849 by J.B. Francis (1815-92), and the Kaplan turbine, designed by Viktor Kaplan (1876-1937). The Pelton wheel, consisting of a ring of buckets or bucket-shaped vanes arranged around the periphery of a wheel, is known as an impulse turbine as it is only the impulse of the water that makes the wheel turn. The Francis turbine, with its outer ring of stationary guide vanes and inner ring of curved vanes on the surface of one side of the wheel, is a reaction turbine; part of the energy is derived from the impulse of the water and part from the reaction between the water and the blades.

The steam-driven turbine was invented in the 1st century AD by Hero of Alexandria. However, the first practical turbine to be driven by steam was a reaction device with several rows of turbinewheels (enabling the energy of the expanding steam to be utilized in stages) invented in 1884 by Sir Charles *Parsons.* An impulse turbine using several steam nozzles was invented by Carl de Laval (1845-1913) in the 1890s. Since the beginning of the 20th century steam turbines based on these designs have repleced the steam engine as the prime mover in *power stations.*

Turbulence. Any random irregularity in the distribution of velocity or pressure in a fluid. Most flows of rivers and winds are turbulent and turbulence in the atmosphere affects aircraft. Turbrulence also causes the laminar layers of flow around an irregular object, or a smooth object at high velocities, to be disturbed, creating a sharp increase in drag. It is to avoid turbulent flow that such objects as aerofoils are streamlined (*see* streamlining; aerodynamics). In some cases turbulence is desirable; in the combustion chamber of a petrol or Diesel engine the designer seeks to create turbulence to improve mixing of the fuel and air.

Turbulent Flow. Flow of a fluid wherein the speed of the fluid change rapidly in magnitude and direction. The motion of a fluid becomes turbulent if its speed increases beyond a certain critical speed. Below this speed, the flow is streamlined.

Turns Ratio. Refrs to the ratio of the number of turns in the primary coil of a transformer to the corresponding number in the secondary coil.

Tweeter. A loudspeaker of small dimension designed to reproduce sounds of relatively high frequency.

Two-phase. Having two equal alternating voltage between which there is relative phase difference of 90°.

U

Ultimate Strength. The value of limiting stress at which a material completely breaks down.

Ultracentrifuge. A high speed rotating machine used for separating very small particles.

Ultra High Frequency. A radio frequency lying in the range of 0.3×10^6 Hz to 3×10^{10} Hz.

Ultramicrobalance. A highly sensitive balance for accurate weighing to 10^{-15} grammes.

Ultramicroscope. A type of microscope, invented by *Zsigmondy* in 1902, used to study colloidal particles in liquid medium. A beam of light iluminates the particles from the side, the scattered light enabling the movements of the particles to be observed as flashes against a dark background (dark-field illumination).

Ultrasonic Frequency. Refers to a frequency above 20,000 Hz.

Ultrasonic Generator. An instrument that produces waves of ultrasonic frequency.

Ultrasonics. The study of *sound waves* inaudible to the human ear, i.e. above about 20 000 hertz. Such waves are known as ultrasound and may be produced by *magnetostriction* or by applying a rapidly alternating voltage across a *piezoelectric* crystal. Ultrasound is used, for example, to destroy bacteria in milk, break up large molecules, and clean surfaces. In medicine, ultrasound is used therapeutically, for example to break up stones, and ultrasound scanners are used in diagnosis and in pregnancy to monitor fetal development, being less harmful than X-rays.

Ultraviolet Rays. Electromagnetic radiation having wave length in the range 50 - 400 A°.

Umbra. The region of compete shadow.

Umklapp Process. A collision process between phonons, or between phonons an electrons. The process is responsible for thermal resistance in non-conducting materials.

Uni- axial Crystal. A double reflecting crystal with only one optic axis.

Unifiler Suspension. A type of suspension used in electrical instruments wherein the moving part is suspended on a signal thread, wire or strip. The controlling torque is formed by the twisting of the thread, etc.

Unified Field Theory. A theory that encompasses the four fundamental interactions—*strong, weak, electromagnetic* and *gravitational*—in terms of a single field, analogous to the electromagnetic or gravitational fields, Einstein failed after many attempts to unify gravtitation and electromagnetism, although the Weinberg-Salam theory achieves unification of the weak and electromagnetic fields.

Unipolar Transistor. A transistor in which current flow is because of the movement of majority carries only.

Unit. A quantity used as standard of measurement in terms of which similar quantities can be measured e.g. In SI system the unit of length is 1 meter, the unit of mass is 1 kilogram and the unit of time is 1 second.

Unit Cell. The smallest crystal unit possessing the entire symmetry of the whole periodic structure.

Unit Planes (and Points). Conjugate planes of a system having a unit lateral magnification.

Unit Pole. A magnetic pole when placed 7 cm from an equal pole, in a vacuum, is repelled with force of 1 dyne.

Unit Vector. Vectors, generally written i, j_0 and k they have unit length and lie along the x y, and z axes respectively. A vector function, F, can therefore be written F = xi + yi + xk. As the angles between these unit vectors are 90°, the scalar and vector products are either zero or one.

Universal Motor. An electric motor that is suitable for use with direct current or alternating current.

Universal Shunt. A galvanometer shunt that is tapped so that it can pass $\frac{1}{10}$, $\frac{1}{100}$, $\frac{1}{1000}$, etc., of the main current through the galvanometer.

Unstable Equilibrium. Refers to a body which on being slightly disturbed from its equilibrium position tends to move away from its original position.

Unsaturated Vapour. A vapour at a certain temperature that does not contain the maximum amount of the susbstance in the gaseous phase.

Upthrust. An upward force that acts on an object immersed in a fluid. The magnitude of the force is equal to the weight of the fluid displaced by it.

V

Vacancy. Refers to an unoccupied site in a crystal lattice that is normally occupied by an atom or an ion.

Vacuum. A region of space that contains no matter. A perfect vacuum is impossible to obtain and any region in which the pressure is less than about 1 mmHg may be considered a vacuum. A soft (or low) vacuum goes down to a pressure of 10^{-2} pascal, a hard (or high) vacuum is between 10^{-2} and 10^{-7} pasca, and an ultrahigh vacuum is below 10^{-7} pascal. Vaccum technology is used in making cathode-ray tubes, light bulbs, etc., and in certain forms of food preservation. *Vacuum gauges* include the *McLeod gauge* and *Pirani gauge.*

Vacuum Evaporation. It is a technique used for producing a coating of one solid on another, usually a coating of metal or semiconductor, by evaporation and high temperature in a vacuum.

Vacuum Ultraviolet. Part of the ultraviolet region of the spectrum, wherein radiation is absorbed by air. The wavelength is less than 200 nm.

Valency Band. Range of energies in a semiconductor that corresponds to energy star that can be occupied by the valency electrons in the crystal.

Valve Reactor. A circuit having a valve that is arranged to operate as a reactor. It is employed in radio transmitters for frequency modulation and in circuits for automatic frequency control.

Valve Voltmeter. Refers to an amplifier that uses thermionic valves with a measuring instrument in the output circuit. Such instruments are used to mesure both d.c. and a.c. voltage.

Vander Waal's Equation of State. An equation of state for real gases,

$$\left(p + \frac{a}{V^2}\right)(V - b) = RT$$

where p = pressure, V = volume, T = absoulte temperature of the gas, a and b are constants, b is a correction term for the actual volume of the gas molecules, which is supposed to be zero.

The $\frac{a}{V^2}$ term arises because of the mutual attraction of the gas molecules.

Vander Waal's Force. Attractive force that exists between the atoms and molecules of all susbstances. It is a short range force changing inversely as seven power of distance. It arises due to molecular dipoles.

Van de Graaff Generator. A type of *electrostatic* generator, invented by the US physicist Robert Jemison Van de Graaft (1901-67), that produces static potentials of millions of volts. Charge from an external source is fed onto a continuously moving belt, which transfers it to the inside of a large hollow coducting sphere. The charge moves to the outer surface of the sphere, leaving the inside neutral and able to collect more charge. This type of generator can be used to provide a high-voltage source for the Van de Graaff *accelerator*.

Vapour. A susbstance in the gaseous state at a particular temperature so that it can be liquefied by the application of a pressure alone. i.e. A gas below its critical temperature.

Vapour Pressure. Vapour pressure of a liquid is the pressure exerted by its vapour at a particular temperature. Its value depends on the nature of the substance and the temperature and is indepedent of the volume of the liquid.

Var. A unit of power similar to watt, used for the reactive power of an alternating current. One var is one volt time one ampere.

Vector. A physical quantity that requires both a magnitude and a direction for its complete description e.g. Velocity, acceleration, force etc.

Vector Product. Of two vectors $\vec{A}$ and $\vec{B}$ is vector $\vec{C}$ written as $\vec{C} = \vec{A} \times \vec{B}$ its magnitude is AB cos θ and its direction is the direction in which a right-handed serew moved from $\vec{A}$ to $\vec{B}$ advance. θ is the angle between $\vec{A}$ and $\vec{B}$.

Vector, Triangle Law of. The resultant of two vector quantities completely represented by two sides of a triangle taken in order is given by the third side of the triangle taken in the reverse order.

Velocity (v). The speed of a body in a specified direction, i.e. its rate of displacement (s) in a specified direction is given by $v = ds/dt$. Velocity is a vector quantity, whereas speed is a *scalar* quantity. In SI units it is measured in metres per second. *Relative velocity* is the velocity of one body relative to another. *Angular velocity* (ω) is the rate of change of angular displacement (θ), i,e $\omega = d\theta /dt$; it is measured in radians per second.

Velocity Ratio. The ratio of the distances through which the points of application of the applied force and the resistance move in a machine. For an ideal machine it is equal to the mechanical advantage.

Velocity, Relative. The rate at which a body, A, changes its position with respect to a second body, B, is known as the relative velocity of A with respect to B.

Venturimeter. A gauge that is used to measure the speed of flow of a liquid. It is based on the Bernulli's equation.

Venturi Tube. A device consisting of an open-ended tube with a central constriction, used to measure the rate of flow of a fluid, which can be calculated from the pressure difference between the centre and the ends. It is used to measure airspeed of aircraft. Invented by G.B. Venturi (1746-1822).

Vernier Scale. A device for measuring subdivisions of a scale such as those on a pair of calipers. The auxiliary Vernier scale is divided so that ten of its subdivisions correspond to nine of those on the main scale. If the reading falls between two main-scale divisions, say between 4.6 and 4.7, the zero mark on the Vernier scale is slid along so that its zero is lined up with the reading. By noting the division on the Vernier scale that is exactly in line with a main-scale division, the second decimal place of the measurement is obtained. If the fourth Vernier division is in line with a main-scale division, the reading would be 4.64. Named after Pierre Vernier (1580-1637).

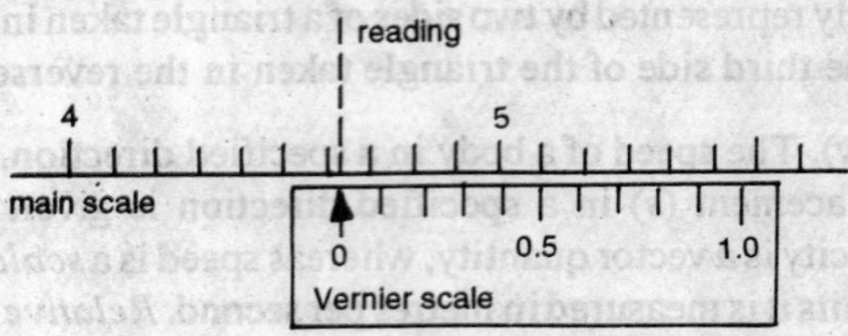

Varnier Scale. **An auxiliary scale used to measure accurately to two places of decimals.**

Very High Frequencies. Radio frequencies between 3×10^7 to 3×10^8 cycles pee second.

Vibration. A periodically repeated to-and for motion about a mean position.

Vibraphone. A type of *xylophone* in which a characterstic vibrato effects is produced by elecrically operated fans installed at the upper ends of the resonator tubes, musical instruments.

Vignetting. Refers to the progressive reduction in the cross-sectional area of a beam of light passing through an optical system as the obliquity of the beam is increased.

Virgin Neutrons. Neutrons from any source which have not been involved in collisions.

Virial Law (due to Clasusius). The mean kinetic energy of a system is equal to its virial that depends only on the forces acting on the atoms and not on their motions.

Virtual Cathode. In a thermionic valve. The surface, situated in the region of a space charge, where the potential is a minimum and the potential gradient (or electric force) is zero. It acts as if it were source of electrons.

Virtual Particle. A short-lived particle that is used in quantum mechanics to represent the interaction between stable particles. For example, in classical physics two electrically charged particles are represented as interacting by the overlapping of the two fields that surround them; in quantum mechanics this interaction would be represented by the exchange of virtual photons between them. The *strong* interaction is represented by the exchange of virtual pions, the *weak* interaction by the exchange of intermediate vector bosons, and the gravitational interaction by the exchange of virtual gravitons.

Violin. An important stringed instrument. It produces musical sound on account of transverse vibrations of the strings. It consists of four strings of steel each with different diameter to produce notes of frequencies 192,288,426.66 and 640 vibrations/sec.

Virtual Work. The work done on a system when it is assumed to be displaced by an infinitesimally small amount. If the system is in equilibrium, the virtual work is zero.

Viscometer (s). Instrument for measuring the viscosity of a fluid. They are based on the properties of fluids that are affected by viscosity.

Viscosity (η). the property of a fluid by virtue of which it tends to resist the relative motion between its different layers. For two successive parallel layers, this frictional force is proportional to the velocity gradient between the layers. The constant of proportionality is known as the coefficient of viscosity of the fluid.

Visible Radiation. Light having wave length in the range between 4000-7000 A^0, that is visible to the human *eye.*

Visual Display Unit (VDU). A *cathode-ray* tube used to display data held in a *computer*. It is usually associated with a keyboard, enabling new data to be entered or existing data to be changed, and sometimes, with a copying device to provide a permanent record or the display.

Vitreous Humour. Gelatinous Substance found behind the lens in the Human eye.

Voiceprint. A graphic record of the sounds produced during speech. The record shows the range of frequencies and harmonics produced and can be used to identify and individual voice. Although there are certain standards in articulation, which are recorded on the voiceprint, individual timbre and harmonics are produced by the shape and flexibility of the larynx and the oral cavity. Voiceprints are used in phonetics and in foensic science.

Volt (V). The *SI* unit of potential, potential difference, or electromotive force equal to the potential difference between two points on a conductor carrying a steady current of one ampere when the power dissipated is one watt. Named after Aleassandro *Volta.*

Voltage (V). The e.m.f. of a source of electricity or potential difference between two points.

Voltage Stabilizer. A device or circuit to maintain a voltage at its output terminals which is the substantially constant and independent of other changes in the input voltage or in the load current.

Voltage Transformer. *Syn.* potential transformer. An instrument transformer wherein the primary winding is connected to the main circuit and the secondary winding is connected to an instrument (e.g. voltmeter). Voltage transformers are widely used to extend the range of a.c. instruments and to isolate instruments form high-voltage circuit.

Voltaic Cell. *Syn.* galvanic cell. A primary cell made up of plates of dissimilar metals dipped in an electolytic solution.

Voltameter. A electrolytic cell where in a metal in deposited on a cathode by electrolysis of a matallic salt.

Volt Amperes. Symbol: VA. the product of a effective voltage and effective amperage in an alternative current circuit.

Voltmeter. A device for measuring the potential difference between any tow points in an electrical circuits. It is a (moving coil) galvanometer having a high resistance connected in series with it. It is always connected in parallel across the desired points in an electrical circuit.

Vortex. The circular motion of a fluid about a vertical axis. Examples include whirlpools, hurricanes and cyclones. A vortex may be formed in a fluid at a point just behind a blunt obstacle past which it is flowing. The direction of rotation of a naturally occurring vortex is often determined by the direction in which the *coriolis* force is acting

W

Wadsworth Prism. An equilaterial glass prism having a plane mirror at 45^0 with the base. A ray passing through the prism at minimum deviation is reflected by the mirror to 90^0 diviation from the incident ray.

Wall Effect. Any significant effect of the inside wall of a container or reaction vessel on the behaviour of the enclosed system.

Wall Energy. Refers to the energy per unit area of the boundary between the domains in a ferromegnetic mateial.

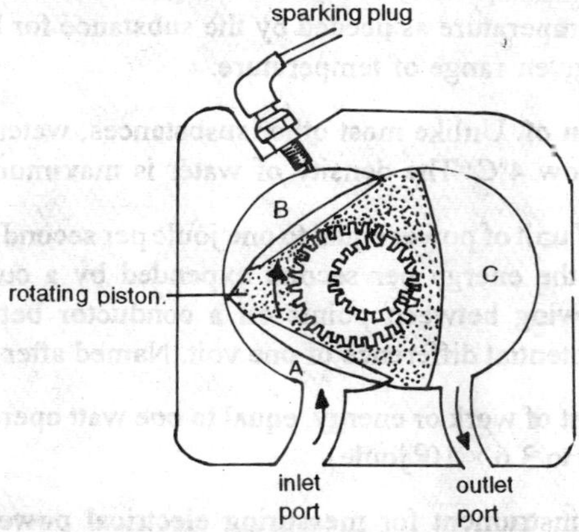

Wankel Engine. The rotary piston draws teh fuel mixture through the inlet port into chamber A. At the same time the gas drawn into chamber B in the previous third of a cycle is compressed and ignited to drive the piston round. Meanwhile, the gas in chamber C is discharged through the outlet port

Wankel Engine. A four- stroke rotary *internal*-combustion engine. It consists of a triangular-shaped rotating piston, with outward curved sides and rounded corners. This rotates in an oval-shaped chamber, slightly narrowed in the middle, which has inlet and exhaust ports and a sparking plug. The piston has an inner toothed annulus, which rotates about a central stationary gear, the whole piston being connected via gears to an output shaft. The small number of moving parts and the lack of vibration are the chief advantages of this type if engine, but gas leakage around the seals between the piston ends and the cylinder has been the principal problem. It is however, used in some cars. It was invented by the German engineer Felix Wankel (1902-88).

Wanner Optical Pyrometer. Polarizing pyrometer.

Warble Tone. A tone in which the frequency changes cyclically between two limits. The frequency change is usually small or compared with the actual frequency of the note and the warble takes place several time per second.

Water Equivalent of a Substance. The amount of water that would require the same quantity of heat for being heated through the same range of temperature as needed by the substance for being heated through a given range of temperature.

Water Expansion of. Unlike most other susbstances, water expands on cooling below 4^{0}C. The density of water is maximum at 4^{0}C.

Watt (W). The *SI* unit of power equal to one joule per second. In electrical terms it is the energy per second expended by a current of one ampere flowing between points on a conductor between which there is a potential difference of one volt. Named after *James Watt.*

Watt-hour. A unit of work or energy, equal to one watt operation for one hour (equal to 3.6×10^3 joule).

Wattmeter. An instrument for measuring electrical power. The most common type has two conducting coils connected in series, one fixed and one movable. The magnetic forces between them produce a deflection of the movable coil proportional to the square of the current, which is in turn proportional to the power.

Wave. Any periodic change in a property of a system that is propagated through a medium (or through space). Waves are classified according to the curve produced when their magnitude is plotted against time on a graph. If the wave is shaped like a sine curve, it is known as a sine (or sinusoidal) wave. Examples of sine waves include electromagnetic waves and sound waves. A wave is characterized by three parameters: amplitude, the maximum displacement of the wave; *wavelength;* and *frequency*. The wave may propagate energy, in which case it is known as a travelling wave, or it may not, when it is known as a standing or stationary wave. For travelling waves the displacement may be either perpendicular to the direction of the propagation of the wave (a transverse wave) or in the direction of propagation (a longitudinal wave). Waves on water and electromagnetic waves are transvese, while sound waves are longitudinal.

Wave Front. The locus of all the points in a wave motion that have same phase.

Waveguide. A hollow conductor for transmitting microwaves over short distances. The electromagnetic radiation is guided along the tube, being reflected from the internal walls. The waveguide is usuall y filled with air but occasionally some other *dielectric* is used.

Wave Intensity. It is proportional to the square of its amplitude. As one moves away from the source of waves, its intensity changes as

$\frac{1}{r^2}$ where r is the distance form the source.

Wavelength. The distance between successive peaks (or troughs) of a wave. It is equal to the velocity of the wave divided by its frequency. For electromagnetic radiation it is a parameter specifying the part of the *electromagnetic* spectrum to which the radiation belongs.

Wave Mechanics. A branch of *quantum theory* in which elementary particles are treated as *de Broglie* waves. Systems of particles are described by a wave equation, known as the *Schrodinger equation*. The solutions of this equation give the allowed values of the energy of each particle (eigenvalues) and the associated wave functions (eigenfunctions); the wave functions provde a measure of the probability that each particle will appear at different points in space. The theory, first proposed by de Broglie, was developed by Erwin Schrodinger.

Wavemeter. A device for measuring radio frequencies. It is made up of an oscillatory circuit tuned to the frequency being measured, generally by means of a variable capacitor that is calibrated in terms of frequency. A device is incorporated that indicates when the circuit has been correctly tuned.

Wave-particle Duality. There is a wave nature associated with moving particles and vice-versa. The wave length (λ) of the wave associated with a particle moving with momentum p is $\lambda = h/p$ where h is Plank's constant.

Wave Motion. Energy transfer by propagation of a periodic disturbance. the disturbance may be in the form of displacement of molecules of a gas, displacement of elements of a string, etc.

Wave Number. The inverse of wavelength, i.e. the number of cycles executed by a wave in a unit length.

Wave Power. The use of the energy of wave motion in the sea to generate electricity. Wave-power generators of various types have been developed, the best known being the "nodding-duck" type, which consists of a string of floats that bob up and down in the waves. The bobbing motion turns a generator. It is estimated that there are between 50 and 100 kilowatts of power per metre in the waves off the coast of Britain, a total of 120 gigawatts over all the suitable sites.

However, only a fraction of this would become electricity. One of the disadvantages of wave power is that it is variable and unpredictable although, like *wind power*, its peak output is likely to coincide with peak demand. Another is that since it is sited off the coast it may be obstructive to shipping and difficult to maintain. In addition, the design and construction of generators and transmission lines to work at sea presents a formidable engineering problem.

Wave Theory of Light. The light energey gets transmitted through space in the form of waves. An imaginary medium either was supposed to fill up the entire space so as to explain the propagation of waves. Later on it was reported experimentally that such a medium does not exist. Light is now known to propagate in the form of electromagnetic waves, that can travel through vacuum as well.

Waves, Audobile. Wave that can be heard by the human ear. The frequency of such waves lies between 20- 20,000 vibration per second.

Wave Train. A successive of waves, particularly a group of wave of limited duration.

Wavetrap. A tuned circuit generally a rejector incorporated in a radio receiver that reduces interference at a particular radio frequency.

Weak Interaction. One of the four fundamental interactions between elementary particles, the others being the *strong, electromagnetic* and *gravitational* interactions. The weak interaction is about 10^{10}times weaker than the electromagnetic interaction. The electroweak theory, proposed by S.L. Glashow (1932-), S. *Weinberg*, and *A. Salam* (sometimes known as the GWS model), unifies the weak and electromagnetic interactions in a *gauge theory*, mediated by photons and intermediate vector boson (the W-boson and the Z-boson).

Weber (Wb). The *SI* unit of magnetic flux equal to the flux linking a circuit of one turn that produces an electromotive force of one volt when reduced uniformly to zero in one second. Named after Wilhelm Weber (1804-91).

Wedge. A piece of metal or wood thick at one end and sloping to thin edge at the other. It is used to separate one part of a body from another or to separate two bodies e.g. Knife, nail chisel.

Weight (W). The force with which the Earth attracts a body. W = *mg* where *m* is the mass of the body and *g*, the acceleration due to gravity. Due to change of *g*, it changes from place to place.

Weightlessness. The condition in which a body possesses no weight. Such a body still possesses mass but its weight can become negligible when the gravitational field is extremely weak, as for example in space. It can also be produced artificially by creating a force that is equal and opposite to gravity. The physiological effects of weightlessness are important in long space flights and are countered by pressure suits, exercise routines, and correct control of cabin pressure, temperature and humidity.

Weston Cell (Cadmium Cell). A standard primary cell used as a standard source of e.m.f. It is made up of a H-shaped glass vessel containing a mercury cathode in one leg and a cadmium amalgam anode in the other. The horizontal part of the vessel contains the electrolyte-a saturated solution of cadmium sulphate. Its e.m.f. is 1.0186 volt at 20°C.

Wet Land Dry Bulb Hygrometer. A device for finding the relative humidity. It is made up of two thermometers placed side by side, the bulb of one of them being surrounded by moistened muslin. The difference in temperatures on the two thermometers enables one to calculate the relative humidity making reference to tables constructed for this purpose.

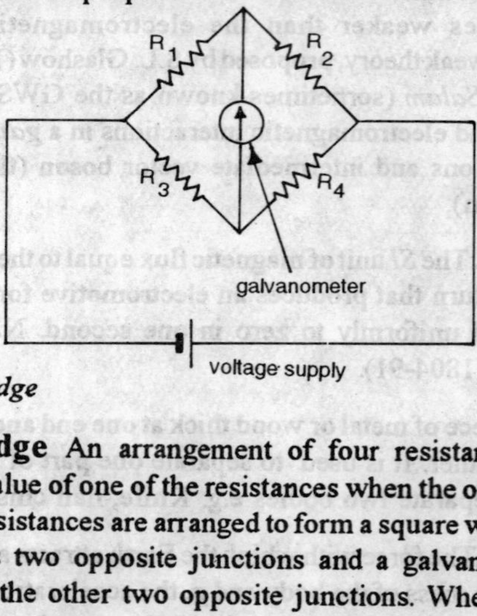

Wheatstone Bridge

Wheatstone Bridge An arrangement of four resistances used to measure the value of one of the resistances when the other three are known. The resistances are arranged to form a square with a voltage applied across two opposite junctions and a galvanometer connected across the other two opposite junctions. When the galvanometer indicates that no current is flowing the bridge is balanced and $R_1 / R_2 = R_3/R_4$.

White Dwarf. Refer to any of a large class of very faint stars that are thought to be in the last stage of stellar evolution. Its nuclear fuel is completely exhausted and it collapses, under its own gravitation, into a small but very dense body.

White Light. Light consisting of a coutinuous range of wave lengths, mainly in the visible range. Such a radiation is given out by an incandescent white hot surface. A mixing of three primary colours produces the sensation of white light.

Wison Effect. When an insulating material is moved through a region of magnetic flux an induced potential difference is set up across the meterial. It becomes electrically polarized this phenomenon is known as the Wilson Effect.

Wimshurst Machine. An electrostatic generator made up of two parallel plates rotating in opposite directions.

Windage Loss. The power loss, generally expressed in watts, that occurs in an electrical machine due to the motion imparted to the gas or vapour (commonly air) surrounding the moving parts, by the latter. This loss is inherent in all electrical machines because ventilation is required for a cooling purposes.

Winding. A winding may consist of a number of separate conductors connected together electrically at their ends or may consist of a single conductor (wire or strip) which has been shaped or bent to form a number of loops or turns.

Wind Instruments. *See* column of air.

Windmills. Machines that enable useful work to be obtained from wind power. The principle is that the energy of the wind turns a set of vanes or sails mounted on a horizontal shaft, the rotation of which is transmitted by gearing to working machinery. Windmills had appeared by 1150 in NE Europe and were used for grinding corn, pumping water, and powering light industry, until they were superseded by steam engines in the 19th centry. Two common designs were the Dutch mill, in which only the sails and the conical roof moved to catch the wind, and the German post mill, in which the whole millhouse with sails attached rotatec around a central supporting pole. The modern metal windmill with multiple-bladed sails is found all over the world in rural areas, pumping water or powering small electric generators.

Window. (1) When a medium, generally opaque selectively transmits a particular small range of radiation it is said to operate as window in that range. (2) The thin sheet of material covering the end of a Geiger counter through which the radiation is received.

Wind Tunnel. A device for testing the flow of air around an object, for example an aerofoil or aircraft, with a view to studying the lift, drag, stremlining, onset of *turbulence,* etc. It consists of duct with an electrically driven fan, usually with water-cooling systems to maintain the air at the correct temperature.

Wobbulator. A signal generator whose output frequency is periodically changed through a definite range or values. It is used as a test instrument for the alignment of tuned circuits and the measurement of the frequency response of electronic circuit and instruments.

Wollaston Prism. A double-image prism cut from a rhomb of calcite. It separats the ordinary and extraordinary components of a ray of unpolarized light. Components tend to show colour effects.

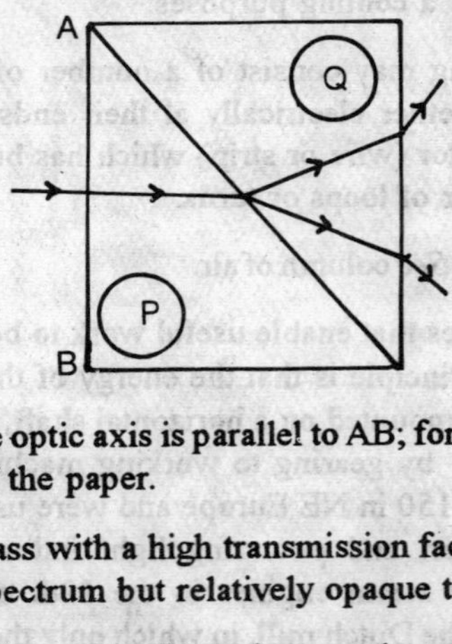

For prism P the optic axis is parallel to AB; for prism Q it is normal to the plane of the paper.

Wood's Glass. A glass with a high transmission factor in the ultraviolet range of the spectrum but relatively opaque to visible radiation.

Woofer. A loudspeaker with dimension, used to reproduce sound of relativley low frequency.

Word. A string of bits used to store information in a computer.

Work. (W). The work done by a force $\vec{F}$ when it moves an object by a distance S is give by $\vec{W} = \vec{F}.\vec{S}$ where $\vec{S}$ is the displacement vector for the object. Thus, $W = FS \cos\theta$ if θ is the angle between $\vec{F}$ and $\vec{S}$. Its units are joule.

Work Function (Φ). The energy that must be supplied to the free electrons in a metal so as to enable them to escape the metal surface.

Working Substance. Refers to the substance that is subjected to temperature and pressure change in a heat engine.

Wow. An undersirable form of frequency modulation heard in the reproduction high-fidelity sound and characterized by changes in pitch below 10 Hz. In the case of a gramophone record it is generally due to non-uniform rotation of the turntable.

W-particle. a hypothetical particle thought to be exchanged between elementary particles undergoing a weak interaction.

Wrench. A force along with a couple whose axis is the line of action of the force.

X

Xerography. A photographic copying process wherein light is passed through the document to be copied and falls on an electrostatically charged plate, generally coated with selenium. This is discharged to an extend which depends on the intensity of the light. A powder with an opposite electric charge is then sprayed on the plate and sticks to the 'dark' areas where the plate has not been discharged. The powder which is a mixture of graphite and thermoplastic resin, is then transferred from the plate to a charged paper where it is fixed by heat treatment.

Xi-particle. Refers to an elementary particle classified as a hyperon.

X-ray Astronomy. The study of celestial objects by the X-rays they emit. As X-rays are absorbed by the earth's atmosphere, observations have to be made above 150 km by rocket or satellite. The first X-ray source, *Scorpius* X-1, was discovered in 1962, since when many sources have been observed, only some of which have been identified optically.

X-ray Crystallography. The study of crystalline susbstances by studying, the diffraction patterns formed by passing X-rays through a crystal. The technique is employed to study structure of crystals as well as that of certain protons and nucleic acid.

X-ray Diffraction. The *diffraction* of *X-rays* when they strike a crystal. The angle through which the X-rays are diffracted depends on the spacing between the different planes in the crystal in a manner given by *Bragg's law.* The technique is used in studying crystal structure.

X-rays. Electromagnetic radiation lying between ultraviolet radiation and gamma rays in the *electromagnetic* spectrum. X-rays may have wavelengths between 10^{-9} metre and 10^{-11} metre, the shorter wavelengths being known as hard X-rays and the longer wavelengths as soft X-rays. Discovered by Wilhelm *Roentgen* (and formerly called Roentgen rays) in 1895, they are produced when heavy metal atoms (usually tungsten) are struck by sufficiently energetic electrons, as in an *X-ray tube*. The electrons in an X-ray tube are produced by a heated cathode in an evacuated tube and accelerated to the heavy-metal anode by an electric field. the collisions knock inner electrons form the atoms, X-rays being emitted when the vacancy is filled by outer electrons. X-rays cause ionization in gases and penetrate matter. X-rays have many uses including the examination of internal organs and structures in medical diagnosis, the killing of cancer cells in *radiotherapy*, investigating structures for flaws, and in *X-ray* diffraction by crystals in order to study their structure.

X-ray Spectrum. Spectrum of X-rays given out by an element when bombarded by cathode rays. The frequencies of the emitted X-rays are characteristic of the atomic number of the element.

Y

Year. It is the time taken by the Earth to complete one rotation around the sun. The *Tropical* or *solar year* is the time taken by sun to successively appear at the first point of Aries and is equal to 365 2422 mean solar days. The civil (or calender year) is regulated with an avarage value of 365.2425 mean solar days. Three successive years consist of 365 days, where as every fourth year (the leap year) is of 366 days. The *sideral* year measured with respect to fixed stars is 365.2564 means solar days, whereas the *lunar year,* equal to 12 lunar months, in 365.3671 mean solar·days.

Yield Point. The point at which a body becomes permanently deformed when subjected to a sufficiently large stress. Below the yield point the body is elastic, above the yield point it becomes plastic.

Yield Stress. The minimum stress for creep to take place.

Yield Value. The minimum value of stress that must be applied to a material so that it shall flow.

Ylem. A hypothetical susbstance made up of neutrons and having a density of about 10^{16} kg m^{-3}. It is supposed to be the material form which all matter has been formed.

Yoke. A fixed piece of ferromagnetic material, that serves to complete a magnetic circuit and that is not surrounded by a winding of any tupe.

Young's Double Slit Experiment. The historic experiment that demonstrated interference of light and gave solid support to wave theory of light. Light from a monochromatic source is allowed to fall on two parallel slits and then on to a screen placed behind. The slits act as two coherent sources of light, from which reaches the screen covering different path. Depending on the path difference between these two, bright and dark regions, known as interference fringes are formed on the screen.

Z

Zener Current. The current in a semiconductor on being subjected to a strong electric field resulting form the escape of electrons from the valence band into the conduction band.

Zener Diode. A semi-conductor diode wherein each side of the junctions is highly doped. When the junction is reverse biased, a sharp increase in the current occurs at a well-defined potential; the current so produced is called zener current and the effect is known as zener breakdown. Used as voltage regulator.

Zero Point Energy. Refers to the enegry possessed by the atoms or molecules of a substance at 0^0K. It cannot be explained on the basis of classical physics but can be accounted for as a quantum effect.

Zeroth Law of Thermodynamics. If two bodies A and B are in thermal equilibrium with a third body C, then A and B are in thermal equilibruim with each other as well.